THE BIG BOOK OF MAKER SKILLS

TOOLS & TECHNIQUES FOR BUILDING GREAT TECH PROJECTS

CHRIS HACKETT AND THE EDITORS OF **POPULAR SCIENCE**

THE BIG BOOK OF MAKER SKILLS

TOOLS & TECHNIQUES FOR BUILDING GREAT TECH PROJECTS

weldon**owen**

CONTENTS

INTRODUCTION from Chris Hackett

BASICS
1. Set Up a Work Space
2. Plan a Project from Start to Finish
3. Play Nicely with Others
4. Keep Track of Tasks for a Team
5. Run a Safe Shop
6. Decide If You Need Stitches
7. Flush Out Your Eyes
8. Rescue a Smashed Foot
9. Save a Finger (or a Toe)

HAND TOOLS
10. Wield a Tape Measure
11. Learn Tape Measure Tricks of the Trade
12. Go with a Combination Square
13. Make Micro Measurements
14. Swing a Hammer Like They Used To
15. Keep Your Hammer in Top-Hitting Form
16. Persuade with Sheer Hammer Force
17. Nail it with a Nailset
18. Pick a Hammer
19. Get Mechanical Advantage from Simple Machines
20. Know Your Fasteners
21. Go Threaded or Unthreaded
22. Decode Screw Heads
23. Sink Some Screws
24. Beat Specialty Screws
25. Deal with a Pesky Stripped Screw Head
26. Put More Torque on a Small Screw
27. Pick a Wrench
28. Improvise a Spanner When You Have No Spanner
29. Boost Leverage with a Cheater Bar
30. Wield a Combination Wrench
31. Get a Grip with a Pipe Wrench
32. Hack Your Pipe Wrench
33. Pick a Saw
34. Focus on Wood
35. Familiarize Yourself with Finishes
36. Master Butt Joints
37. Sand It Right
38. Shear and Saw
39. Use a Tube Cutter
40. Choose Sheet Metal Snips
41. Know Your Chisels
42. Pick a Vise, Clamp, or Jig
43. Use an Old-School Tool for a Manufactured Effect
44. Tap a Hole Like a Pro
45. Thread a Pipe with a Die
46. Set Up a Gas Welding Station
47. Protect Your Floor from Sparks
48. Check Gas Tank Hoses for Leaks
49. Make Your First Oxyacetylene Weld
50. Braze with a Torch
51. Bend Metal with Fire
52. Build Your Own Welding Table
53. Cut Heavy Metal with Fire
54. Flame-Cut in Straight Lines
55. Go on the Hunt for Obtainium
56. Assemble a Paint-Can Forge
57. Forge a Knife
58. Cast a Shot Glass
59. Transform a Forge into a Foundry
60. Craft a Crucible (or Two)
61. Fashion a Crucible Carrier
62. Assemble a Crucible Pourer

63	Hammer a Dross Skimmer	82	Know Your Stitches
64	Refine Aluminum	83	Create Quick and Dirty Patterns
65	Focus on Masonry	84	Win at Life with No-Sew Fasteners
66	Mix a Batch of Concrete		
67	Make a Solid Base for a Structure		
68	Butter a Brick		
69	Cast a Plaster Mask		

POWER TOOLS

70	Set Up a Glass Work Space		
71	Mount Glass	85	Pick a Power Saw
72	Cut Glass Bottles with String	86	Make Masterful Curves with a Jigsaw
73	Pick Pliers	87	Choose a Circular Saw Blade
74	Tackle Tapes	88	Slice with a Circular Saw
75	Choose an Adhesive	89	Work with a Portable Bandsaw
76	Lock (and Unlock) That Thread	90	Bench-Mount a Portable Bandsaw
77	Focus on Paper & Cardboard	91	Get Inventive with a Bandsaw
78	Make a Screen for Screen Printing	92	Respect the Angle Grinder
79	Burn an Image onto a Screen	93	Pick a Sander
80	Screen-print by Hand	94	Do It All with an Angle Grinder
81	Assemble a Maker's Sewing Kit	95	Pick a Wheel, Any Wheel

96	Use a Cut-Off Wheel	118	Etch Your Own Circuits
97	Drill Better Holes	119	Pick Electronic Components
98	Drill Glass Without a Crack	120	Know Your LEDs
99	Handle a Hole Saw	121	Wire the Simplest Circuit
100	Chain-Drill Slots	122	Store Energy with a Capacitor
101	Pick a Drill Bit	123	Regulate with a Potentiometer
102	Know Your Rotary Tools	124	Direct Current with a Relay
103	Transform a Rotary Tool as a Tiny Drill Press	125	Calculate LED Resistor Needs
104	Focus on Metal	126	Pick Illumination Sources
105	Understand Electrical Welding	127	Fade an LED
106	Set Up a Space for Electrical Welding	128	Wire in Series or in Parallel
107	Start with Stick Welding	129	Work with High-Power LEDs
108	MIG Weld an Inside Corner	130	Master Multiplexing
109	Try Some TIG Welding, Too	131	Get the 411 on the 555
110	Weld Like a Badass	132	Set Up a Free-Running Oscillator
111	Hack Together a Welder	133	Wire a One-Shot Timer
112	Fuse Metal with Your DIY Welder	134	Meet the 555 Family
113	Set Up a Soldering Station	135	Get to Know Your Multimeter
114	Solder to a PCB	136	Check for Continuity
115	Solder Wires Together	137	Measure Voltage
116	Decode Circuitry Schematics	138	Gauge Power
117	Prototype with a Breadboard	139	Sew Circuits on Fabrics

140	Insulate Your Conductive Thread
141	Pick Fabric for Its Resistance
142	Focus on Textiles
143	Experiment with Flexible Switches and Sensors
144	Use a Multimeter to Check Material Resistance
145	Sew a Simple Soft Circuit Cuff
146	Know Your Photodetectors
147	Improvise a Pressure Sensor
148	Build a Loop Switch
149	Decipher Decibels
150	Get Various Noises from a Buzzer
151	Work with Speakers
152	Build a Basic DIY Amp
153	Master Microphones
154	Explore Sounds with a Piezo Mic
155	Bend That Circuit
156	Get to Know Radio
157	Listen in with a Trench Radio
158	Rig a Cell-Phone Blocker
159	Give a Homopolar Motor a Spin
160	Wire Up a Reverse Switch
161	MacGyver a Generator
162	Strip a Drill for Parts
163	Roll Your Own Gears
164	Shim a Gearbox
165	Pick a Power Source
166	Harvest Electronic Obtainium
167	Cook with the Sun
168	Get Solar Panel Savvy
169	Save Daylight in a Jar
170	Scheme Up a Wind Turbine
171	Power Up with Pedal Power
172	Build a Bicycle Generator

ROBOTS & BEYOND

173	Meet Your Robot Minions
174	Scope Out the Software
175	Get Access to High-End Tools
176	Understand 3D Printing

177	Find a 3D Printer Near You
178	Prepare to Print a 3D Model
179	Enter the New Age of Replicators
180	Focus on Plastic
181	Pick a Microcontroller
182	Accessorize Your Microcontroller
183	Make a Case for Prototyping
184	Use Programming to Blink an LED
185	Pick an Input or Output
186	Take a Superquick Programming Primer
187	Build an Add-On Board
188	Hail the Lasercutter
189	Lasercut the Right Materials
190	Make Your First Lasercut Design
191	Troubleshoot a Lasercutting Job
192	Set Up a Hackerspace
193	Pick a Crowdfunding Platform
194	Put Together a Pitch Video
195	Build Community
196	Learn the Basics of Robot Anatomy
197	Build a Simple Proximity Detector
198	Make Your Robot See
199	Give Your 'Bot Touch Sense with Guitar String
200	Help Your 'Bot with Distance Detection
201	Hack a Servo for Continuous Rotation
202	Clown Around with a Balloon Gripper

Glossary
Index
Acknowledgments & Credits

INTRODUCTION

I was one of those children always taking things apart—common to makers, but looking back, I realize that child me was a lot weirder and darker than one would want: I took stuff apart, but with zero interest toward improving the objects or even putting them back together. As I recall, I had no curiosity as to what the little bits did, and the concept of a "soldering iron" was totally foreign. I just wanted to carefully, methodically destroy things. I think we are all glad I did not have easy access to frogs or puppies.

But as I got older, I started to develop an interest in how the world worked, and realized my ignorance was deep and vast. I paid my electric bills, but where did the electricity come from (wait, back up—what is electricity?). And what, exactly, was it doing once it got here? I wanted to create things, but I was clueless as to where to begin or even the correct terms to use. Like more and more of us, I grew up in a world that downplays the physical and the practical—where the vague menace of lawsuits makes shop classes disappear and distaste toward working with your hands leaves us all idiots staring at screens.

Slowly I learned the words (thanks, McMaster-Carr) and then the techniques. A great thing about "fake it until you make it" is that, once you've made it, you no longer have to fake it. The first things you make will look like crap, will probably not work, and will become cherished, deeply confusing family heirlooms. That's OK. In fact, that's all part of it.

This book is designed for the people starting from zero, the people who know a little and want to know more, and those who are pretty good and want some neat tricks. Writing it was harder than I thought it would be, but I hope it allows you to have hands that never come totally clean and a deep appreciation for the wonderful stuff that holds our world together.

CHRIS HACKETT

BASICS

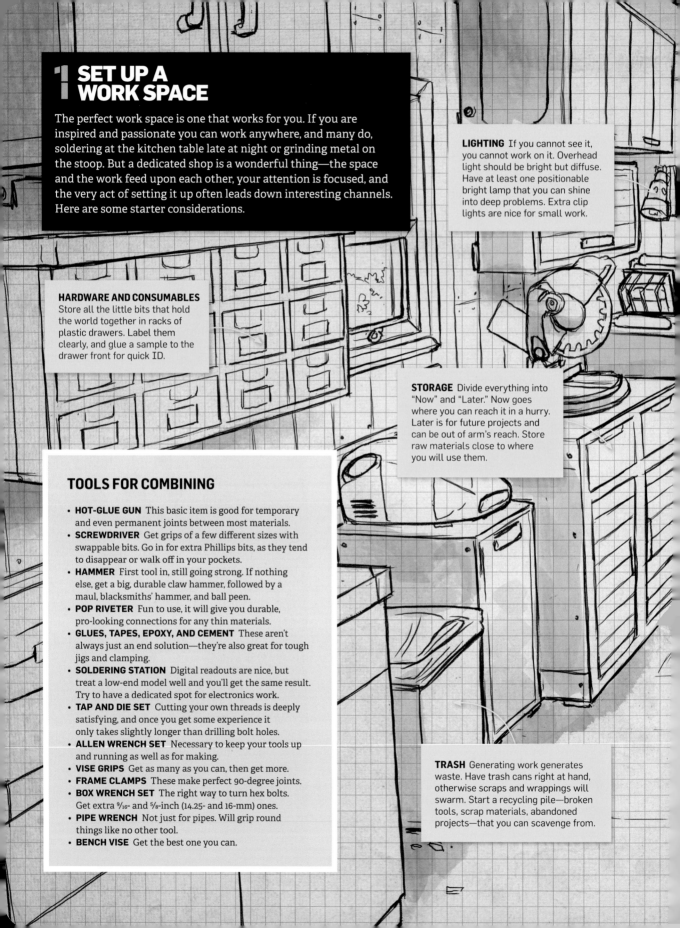

1 SET UP A WORK SPACE

The perfect work space is one that works for you. If you are inspired and passionate you can work anywhere, and many do, soldering at the kitchen table late at night or grinding metal on the stoop. But a dedicated shop is a wonderful thing—the space and the work feed upon each other, your attention is focused, and the very act of setting it up often leads down interesting channels. Here are some starter considerations.

LIGHTING If you cannot see it, you cannot work on it. Overhead light should be bright but diffuse. Have at least one positionable bright lamp that you can shine into deep problems. Extra clip lights are nice for small work.

HARDWARE AND CONSUMABLES Store all the little bits that hold the world together in racks of plastic drawers. Label them clearly, and glue a sample to the drawer front for quick ID.

STORAGE Divide everything into "Now" and "Later." Now goes where you can reach it in a hurry. Later is for future projects and can be out of arm's reach. Store raw materials close to where you will use them.

TOOLS FOR COMBINING

- **HOT-GLUE GUN** This basic item is good for temporary and even permanent joints between most materials.
- **SCREWDRIVER** Get grips of a few different sizes with swappable bits. Go in for extra Phillips bits, as they tend to disappear or walk off in your pockets.
- **HAMMER** First tool in, still going strong. If nothing else, get a big, durable claw hammer, followed by a maul, blacksmiths' hammer, and ball peen.
- **POP RIVETER** Fun to use, it will give you durable, pro-looking connections for any thin materials.
- **GLUES, TAPES, EPOXY, AND CEMENT** These aren't always just an end solution—they're also great for tough jigs and clamping.
- **SOLDERING STATION** Digital readouts are nice, but treat a low-end model well and you'll get the same result. Try to have a dedicated spot for electronics work.
- **TAP AND DIE SET** Cutting your own threads is deeply satisfying, and once you get some experience it only takes slightly longer than drilling bolt holes.
- **ALLEN WRENCH SET** Necessary to keep your tools up and running as well as for making.
- **VISE GRIPS** Get as many as you can, then get more.
- **FRAME CLAMPS** These make perfect 90-degree joints.
- **BOX WRENCH SET** The right way to turn hex bolts. Get extra $9/16$- and $5/8$-inch (14.25- and 16-mm) ones.
- **PIPE WRENCH** Not just for pipes. Will grip round things like no other tool.
- **BENCH VISE** Get the best one you can.

TRASH Generating work generates waste. Have trash cans right at hand, otherwise scraps and wrappings will swarm. Start a recycling pile—broken tools, scrap materials, abandoned projects—that you can scavenge from.

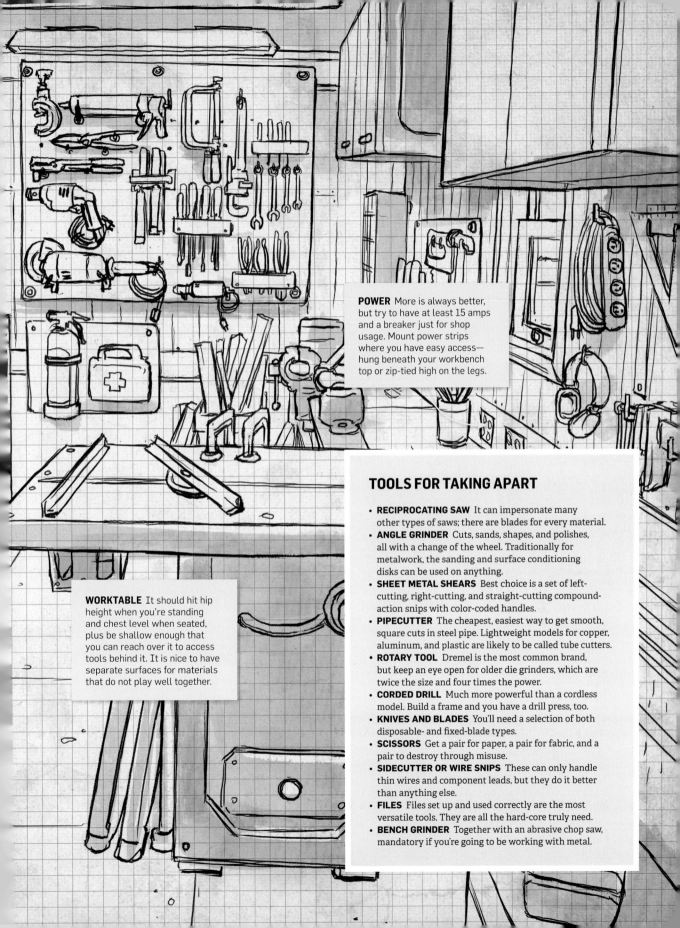

POWER More is always better, but try to have at least 15 amps and a breaker just for shop usage. Mount power strips where you have easy access—hung beneath your workbench top or zip-tied high on the legs.

WORKTABLE It should hit hip height when you're standing and chest level when seated, plus be shallow enough that you can reach over it to access tools behind it. It is nice to have separate surfaces for materials that do not play well together.

TOOLS FOR TAKING APART

- **RECIPROCATING SAW** It can impersonate many other types of saws; there are blades for every material.
- **ANGLE GRINDER** Cuts, sands, shapes, and polishes, all with a change of the wheel. Traditionally for metalwork, the sanding and surface conditioning disks can be used on anything.
- **SHEET METAL SHEARS** Best choice is a set of left-cutting, right-cutting, and straight-cutting compound-action snips with color-coded handles.
- **PIPECUTTER** The cheapest, easiest way to get smooth, square cuts in steel pipe. Lightweight models for copper, aluminum, and plastic are likely to be called tube cutters.
- **ROTARY TOOL** Dremel is the most common brand, but keep an eye open for older die grinders, which are twice the size and four times the power.
- **CORDED DRILL** Much more powerful than a cordless model. Build a frame and you have a drill press, too.
- **KNIVES AND BLADES** You'll need a selection of both disposable- and fixed-blade types.
- **SCISSORS** Get a pair for paper, a pair for fabric, and a pair to destroy through misuse.
- **SIDECUTTER OR WIRE SNIPS** These can only handle thin wires and component leads, but they do it better than anything else.
- **FILES** Files set up and used correctly are the most versatile tools. They are all the hard-core truly need.
- **BENCH GRINDER** Together with an abrasive chop saw, mandatory if you're going to be working with metal.

2 PLAN A PROJECT FROM START TO FINISH

It's been said that Rome wasn't built in a day—and the Romans had nearly endless resources and conscripted labor. While maker projects tend to be on a smaller scale than Mediterranean empires, a little bit of forethought still goes a long way.

DO YOUR HOMEWORK Stand on the shoulders of the maker giants before you. Check the Internet (or even the library), and pick the brains of pals or experts who might have completed similar projects in the past. There's no need to make mistakes others have already made for you.

SKETCH Before you start swinging a hammer, sketch out your plans. If you want to get fancy, you might even try to capture your vision using 3D modeling software.

OBTAIN TOOLS AND MATERIALS Procure your materials—the more salvaged, the cheaper and the better. If you're missing an important tool, now would be the time to call in a favor with that neighbor who owes you one. If that doesn't work, try a tool library, and a hardware store if you must.

DIVIDE AND CONQUER Based on your research, sketches, and available tools and materials, portion out the project into smaller parts. It can be a lot less daunting to think about tackling a project piece by piece than as a whole.

JUST DO IT With a plan in place, all that remains is, you know, actually following through. Start at the beginning. Work until finished. Easier said than done.

3 PLAY NICELY WITH OTHERS

As a maker, doing things on your own might come more naturally than working in groups. Still, if you hope to make on a scale larger than you can hope to do independently, it helps to have a crew. Of course, that means you have to get along with people who aren't you. Here are a few tips for doing so, from one angry loner to another.

PICK A TEAM When making with a crew, you don't necessarily want a bunch of additional yous on the team. In fact, aligning yourself with people who have skills, contacts, and resources that you lack can be a really big help. Building a tiny house but only know carpentry? Sounds like you might be on the lookout for a plumber and an electrician. Tricking out a car with neon and a booming sound system but only know lighting? Sounds like you need an audiophile.

TALK Once you've found people whose skills complement your own, the challenge becomes getting everyone on the same page. Different personalities may require different communication styles. Talk through ways that work for everyone at the outset.

KEEP THE PEACE If at any point things start to go sour, address it in a way that is open, respectful, and direct. A hairline fault, ignored too long, can grow into a crevasse of resentment within your team. It's a lot easier to patch a small crack than build a bridge.

BRIBE When all else fails, a beer can sometimes help smooth out the rough patches. Just be sure the power tools are powered down before imbibing.

4 KEEP TRACK OF TASKS FOR A TEAM

Little personal projects—the stuff that most people can keep updated in their brain or on a handy app—require little lists. Add complexity—systems instead of parts, helpers requiring guidance, looming deadlines—and the system breaks. One day, your project manager will run out for a pack of cigarettes and never, ever return.

There is a better solution: giant pieces of cardboard. Boldly write lists of work that needs doing, in order, with a highly visible check box next to each task. There are few things as satisfying as checking off a box.

PUT IT IN ORDER Organize tasks into a logical sequence—steel needs to be measured, marked, and cut before holes get drilled; it's a lot easier to drill through loose parts, so assembly should go after drilling whenever possible..

PUT SIMULTANEOUS TASKS TOGETHER If two sets of tasks can be done at the same time, break the list into columns to reflect that.

MAKE IT VISIBLE Hang specific lists near the machine or work area they apply to (cut lists next to the saw, drill lists on the drill press, etc.). An even larger list of lists—the whole of the project in paper form—should be visible from all points. With progress and bottlenecks posted in clear sight of all, deadline problems cannot take anyone by surprise.

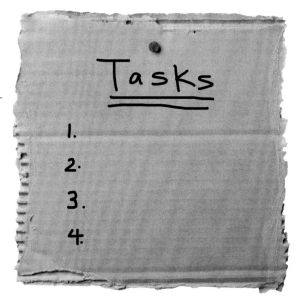

5 RUN A SAFE SHOP

Getting into making shows us the world in a new light—once you have done metal fabrication, you find yourself silently critiquing welds, and designing a circuit gives you a new appreciation of consumer electronics. You also discover some really interesting ways to be horribly, horribly injured. Even if you don't go looking for trouble, it will find you, especially when you're tired or impatient. Be prepared by having extra protective gear all around your work space, and make proper use of that equipment a standard operating procedure.

SAFETY GLASSES FOR ALL This can't be said often (or loudly) enough: All shops want to blind you. Metalwork is all about ejecting tiny chunks of razor-sharp, red-hot stuff at high speed. Tear glands packed with fine sawdust are painful and hard to wash out. Touch solder paste, then touch your eye, and you will know pain. Nasty vapors from etching tanks have left me with red eyes for a week. Put a bin of safety glasses by the door and post a notice that everyone must wear them at all times. No exceptions.

VENTILATE (OR GET OUT) For most tasks, it's enough to open a window or door and run a fan. Pro-level ventilation is pricey, but a cheap alternative is to simply work outside. At my shop, spray-painting, cleaning metals with acid, and sanding wood take place in the backyard, no matter how cold it is or how much time it adds.

HAVE A FIRST AID KIT Get an OSHA-approved one, then add in extra bits, like a small cooler and instant-cold packs (in case any body parts get loose and need to be packed for the ER), what will seem like way too much gauze, a couple of tubes of superglue, a surgical staple kit, and a stack of suture kits. **Pro tip:** Keep the first aid kit close to the ground so you can reach it even if you can't stand upright, and hang a big red sign so everyone knows where it is. They might get to use it on unconscious you.

CHOOSE YOUR HEARING BATTLES Ear protection is the most uncomfortable, therefore the most shunned, bit of safety gear. I never wear it—I like to listen to the shop to make sure everything is working correctly. But certain tools, like high-speed electric motors and large angle grinders, operate naturally at a high pitch that you can feel as well as hear. That feeling is pain, and the pain is parts of your inner ear dying. Stock quality earplugs and Velcro earmuffs near these tools and put them on.

YELL WARNINGS If you work in a space with other people, the thing you are armored against might affect the person a few work stations away. Around here, it is courteous to yell "LOUD!" before firing up the chop saw or impact wrench or "WELDING!" before striking an arc (sometimes, I yell "WHOSE KEYS ARE THESE?"). A moment's notice is all people need to put hands over ears or avert eyes, so be sure to give them at least that.

6 DECIDE IF YOU NEED STITCHES

There are many, many tools that cut, and their intended targets are usually a lot tougher than you are. Still, in my experience, the things thirsty for blood are not the tools but the other stuff around the shop: screws just a little too long for the wood, fresh burrs from drilling, or a too-hard grab at a falling beer bottle. Here's what to do if you can see into your soul, and it is bleeding.

STEP 1 Can you see bone, or is the wound on a flexing part? If no, proceed to step 2; if yes, see step 5.

STEP 2 Apply direct pressure—sterile gauze is best, clean towels or fresh laundry are OK, and your shirt works if nothing else is available. Did the bleeding stop? If yes, proceed to step 3; if not, see step 5.

STEP 3 Clean the wound. A saline flush is best, but tap or bottled water will work.

STEP 4 Evaluate the situation. Will a bandage or wrap hold the sides of the wound together easily? If yes, wrap it and keep it clean; if no, see step 5. If the wound will not close because it's too torn,

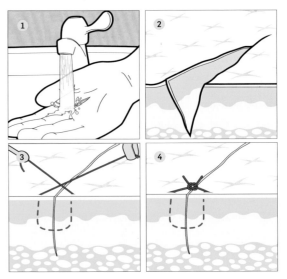

or parts are missing, wrap it and keep it clean and again, see step 5. **Pro tip:** Superglue will bond most edges together better than thread. Actual doctors do this. Trust me.

STEP 5 If all efforts at basic bandaging fail, head to the emergency room— or try your hand at stitching it up yourself, shown above.

WORK IT GOOD

7 FLUSH OUT YOUR EYES

At this point, I have no idea if the slogan "Safety Third!" is a joke or not. (Repeated concussions will do that to you.) Minor injuries are a part of making, but getting something in your eye is no fun at all.

STEP 1 Immediately flush the eye using eye cups and emergency eye wash. If you wait, you will rub, gouging racing stripes into your cornea. If you do not have eye cups, pry the injured eye open with one hand, and use the other to squirt saline directly at the ouchie. **Note:** This is easier with another person—you hold, they flush. No saline? Ask around. Ten percent of Americans wear contact lenses, but anti-redness eyedrops will work and water will do in a pinch. Disinfected shot glasses also work well as eye cups.

STEP 2 Eye still hurt? Look at it, or have someone look for you. If you do not see an object, your eye is just irritated. Give that eye a break—wear a patch (no, I am not joking). See something inside your eye? Proceed to step 3.

STEP 3 If you see steel, take a strong rare-earth magnet and superglue it to a stick. Hold your eye open and pass the magnet as close to the metal as possible; the magnet will remove the metal. If the object was nonferrous, you can wait, flushing every hour or so, and it will be expelled or absorbed in a few days. Object won't flush out or you want to avoid a few painful days? See step 4.

STEP 4 Take your hurt peeper to see a doctor.

8 RESCUE A SMASHED FOOT

The maker world is full of heavy things, dense things, things that have awkward angles and bizarre balances—in other words, things that want to drop. Sometimes, those things get their wish. Between your hands and the center of the earth are your feet, and sometimes the falling object comes to rest on top of you.

STEP 1 Stop screaming. Sit down. Remove your shoe and sock.

STEP 2 Do you see normal toes that are rapidly getting less normal—either swelling or topped with a toenail that's racing to turn black? Your toe is probably broken. Ice it, elevate it, stay off it. If it hurts, consider it the price of getting all the benefits of gravity. If the pain kicks up tremendously when you walk or move, do what's called a buddy splint: Tape your broken toe to the unbroken toe next to it. Broken toes take about six weeks to heal. If the pain ramps up to excruciating, then proceed to step 5.

STEP 3 See neatly severed toes, bones sticking out, some kind of pulp that looks like canned dog food, or nothing, because everything came off with the sock? See normal toes, but bent the wrong way, compressed into a negative of whatever hit them? See step 5.

STEP 4 Hobble to an ER to get your foot fixed as quickly as you can.

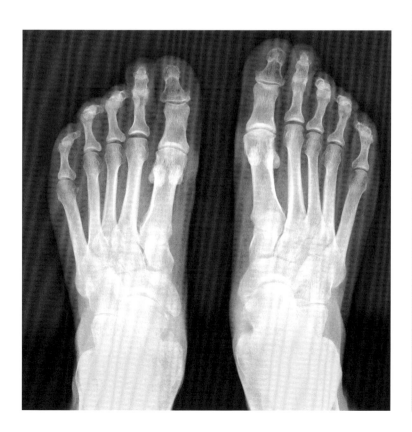

9 SAVE A FINGER (OR A TOE)

Makers live in a world of spinning sharp things, or tools from the good old days, when America prided itself on its quality manufacturing and "safety" was purely a football term. Some of our creations are a little on the risky side, and when a maker modifies a tool or piece of equipment, the safety features are the first to go. The second thing to go is a finger or a hand—feet and legs, more rarely.

STEP 1 First things first: Try to stop the blood using direct pressure. If you bleed through a bandage, add another.

STEP 2 Find the part or parts. Pick it up with a clean, damp cloth.

STEP 3 Wrap the cloth and severed part in clean plastic. Make sure there is a little bit of air inside the plastic—vacuum-sealing keeps in the flavor but kills the part. You want the part alive.

STEP 4 Put the wrapped part in a cooler or a bucket of water with some ice—your goal is to keep the part cold but not freezing. Also, do not get it wet.

STEP 5 Go to the emergency room as fast as you can. You have maybe 4 to 6 hours for any chance of reattachment.

HAND TOOLS

10 WIELD A TAPE MEASURE

I often say, "Build it, then measure it." This is horrible advice—and an excellent way to waste time making parts that could have fit together and materials that could have been put to good use. At the same time, overly precise measurements can cause unneeded frustration and waste time, too, since different tasks require different levels of precision. When it comes to measuring anything larger than your torso, the tape measure will get you in the ballpark. Putting it to use is pretty obvious stuff: Pull out the end as far as you need, fix the reel in place with the lock, and then measure as precisely as you can against the markings on the tape.

But there are a few neat features that might not be so obvious. The curve of the tape, for instance, has some structural rigidity to it, allowing you to measure higher and across gaps that you can't physically reach across. Just play the tape up or out, bit by bit. Also, most tapes can bend to measure around curves or corners, but take care to orient the concave surface toward the inside of the bend. Otherwise you might kink or crack the tape.

"Always remember that there are many, many tools that make things shorter—and achingly few that make things longer."

11 LEARN TAPE MEASURE TRICKS OF THE TRADE

PROPER RETRACTION Do not retract the tape by just releasing the lock and letting the tape slam home. The shock of impact loosens the lip, weakens the spring, and makes you look like an amateur. Instead, release the lock while keeping a finger or thumb on the tape, and let the spring pull it back slowly and smoothly.

TRUE ZERO The end bit (called the "tang" or "lip") of a tape measure comes loose in normal use—do not trust it to be at zero. Rather, measure from the 1-inch (2.5-cm) line, and adjust your measurements accordingly.

GET MORE LENGTH The length of the tape measure's body is usually written on or embossed on the bottom. Add this to the measured distance to get a little more reach out of the tape, or to measure in tight corners.

12 GO WITH A COMBINATION SQUARE

A combination square helps measure smallish things (from the size of your torso down to your thumb) and is also handy for checking angles, marking parallel lines, and finding centers. You can spot this common, accurate layout tool by its two parts: a steel ruler and one of several sliding heads. The ruler usually has four different scales (one for each long edge, on both sides) and the most common standard head includes 90- and 45-degree faces, plus (often) a bubble level and hard steel scribe you can pull out to mark stuff. Loosen and tighten the nut as needed to fix the head where you want it along the ruler.

To measure depth, lay the right-angle face across the opening of a hole and run the ruler down to the bottom; to measure height, use the face as a base and run the ruler up. Align the face to an object's edge and use the long side of the ruler to check perpendicularity, or the short side to check parallelism (which can be handy, for instance, when you need to check the trueness of a box or frame). Fancier combination square sets include a protractor head, which is adjustable to angles other than 90 and 45 degrees, or a center square head, which arranges one edge of the ruler to bisect a precise 90-degree angle machined into the head and allows you to quickly and accurately find the centers of round objects.

WORK IT GOOD

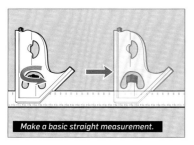
Make a basic straight measurement.

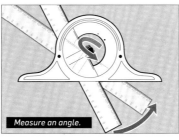
Measure an angle.

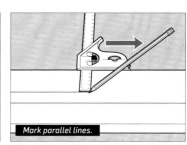
Mark parallel lines.

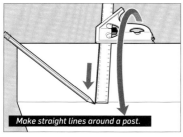
Make straight lines around a post.

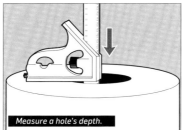
Measure a hole's depth.

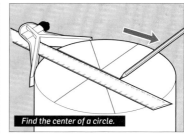
Find the center of a circle.

13 MAKE MICRO MEASUREMENTS

The caliper is for measuring small things, from the size of your thumb down to a hair, all to a satisfying degree of accuracy. Most calipers can measure to the thousandth of an inch, high-end ones to the millionth. Until recently, digital calipers were an expensive specialty tool—the type that the owner kept safe and never, ever lent out. Today, cheap ones pop up in big-box hardware stores, but they don't have millionth-of-an-inch accuracy, and they tend to die if they get wet or fall off a workbench. Even so, they're fine for most jobs, and I recommend a quality (~US$150) digital caliper or a decent dial caliper—the batteries will never die, and analog accuracy was good enough to split the atom and land men on the moon.

To measure with a caliper, make sure the display is at zero when the caliper is closed—if not, re-zero with the calibration knob or button. Use the large main jaws to take outside measurements by straddling a part or distance. Use the smaller jaws to measure an inside gap or hole. The rail that comes out of the opposite end can be used as a depth gauge. Tighten the lock knob to hold the jaws in place when comparing sizes or using the jaws as a scribe.

14 SWING A HAMMER LIKE THEY USED TO

Take a long look at this old workhorse, and you'll remember something that you've known along: This thing's really just a lever. It's got two parts: a handle (which extends the length of your arm and the arc of your swing) and a head, a dense plane that focuses your energy on a point of impact. This creates a mechanical advantage that can drive a nail into wood, distort metal into a new shape, or just smash what needs smashing. To use it:

STEP 1 Firmly grip the handle about one-third of the way from the bottom, thumb toward the head, with the handle perpendicular to your forearm. Make sure you've got the head pointing the right way.

STEP 2 Take a practice swing. Hold the head high (near your own), bend your arm at the elbow, rotate your shoulder back a little, and then bring your arm through the full range of motion, making a smooth arc to the point of impact. Make light contact, moving slowly, and keep your eyes on the target, not the tool.

STEP 3 Time for a real swing. This time, make a quick and full-force arc. At the end of the stroke, almost all the energy stored in the moving head should dump into the target. Hold back just enough back to maintain form and repeat as needed.

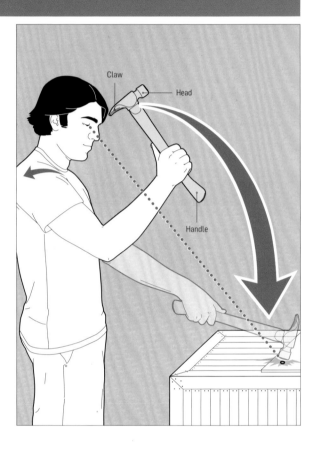

15 KEEP YOUR HAMMER IN TOP-HITTING FORM

Hammers are the gold standard for tough, but bad habits or sloppy maintenance can cause one to fail in ways that are at best inconvenient and at worst dangerous. These guidelines will save you from breaking the unbreakable.

MIND THE GAP The weakest part of any object is usually the joint; on a hammer, this is where the handle passes through the head. If things jostle, the head can turn into a lever-launched projectile. Fix any looseness by adding new wedges or driving current ones deeper.

COVER THE NECK Swing a hammer enough and you will occasionally misjudge the distance and smash its handle. Protect yours by wrapping it in electrical tape.

GO EASY ON THE CLAW Some like to sleeve the hammer's handle with a piece of pipe (see #29) to increase its leverage when prying with the claw. This is a great way to snap the claw right off. Instead, hit the face of the hammer with another hammer, driving the taper of the claw under the obstacle to pull it steadily up and out.

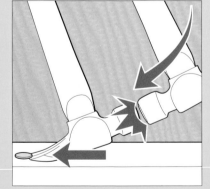

HIT SQUARELY Strive for ideal hammering form, in which the front face of the head contacts the target at a perfect 90-degree angle. Glancing blows or, worse still, trying to use the side faces instead of the front, can result in sloppy, damaged work or, in extreme cases, a broken tool.

16. PERSUADE WITH SHEER HAMMER FORCE

Spend any time in the world of making and eventually you'll need to move very large things: welding tables, whole walls, equipment made in an era when the default engineering solution was to add another ton of cast iron, cast-off infernal machines made of purest heavy. Long levers snap, combined shoulders do nothing. Instead of despairing and pricing forklift rentals, use a hammer.

I know this technique as "persuading," one of those terms that is immediately understood as soon as you say it while gesturing with a hammer. If you have an immovable object, give it a good, hard thump, below the center of gravity, parallel to the ground, in the direction you want it to go. Use the largest hammer available. Move the hammer slowly—think croquet, not slap shot—and steadily. Follow up the thump with a shoulder, and chances are the immovable object will get moving.

Why does this work when a lever fails? Moving a huge object requires a large impulse to overcome its static friction with the ground. Impulse depends on the total force and the time over which it is applied. A force applied over a short time, like a hammerstrike, can deliver a larger impulse than the same force over a longer time.

> "Our ancient ancestors could not speak and were noncommittal on bipedalism...but they did have hammers, and they swung them just like you."

17. NAIL IT WITH A NAILSET

Hammers and nails coevolved, meant for one another. When the heavy face of the hammer hits the head of a nail, there's only one place for the force to go—down through the nail shaft to its sharpened tip. It's a lever driving a wedge, mechanical advantage piled on mechanical advantage, a minor miracle of ingenuity. But before you get too excited, remember to ease up as the nailhead gets closer to the work surface. One overzealous swing, and the nail gets buried in the center of a hammer-shaped divot, marring what could've been a nice finish job.

The solution? Use a nailset, which is a punch that goes between the hammer and the head of the nail to channel the driving force during those last few critical blows. Just place the small end against the nailhead with your nondominant hand, hold it steady, and tap the large end with your hammer until the nail sits flush with the wood.

18 PICK A HAMMER

Normally, feats like breaking rocks or distorting metal are the domain of ninjas and circus strongmen, but simple, force-concentrating tools like hammers turn us all into death-dealing circus freaks. The lever arm of the handle makes the head move faster than your hand, and the relatively high mass of the head stores all of the energy until it gets dumped into whatever you're whacking.

CLAW HAMMER Classic, everyday hammer used primarily for pounding nails into and pulling nails out of some object.

MALLET Used to deliver a good amount of force without damaging the surface or object you're striking. Rubber, wood, and brass are the most common materials for the heads.

MAUL Some are like mini sledgehammers used for driving wedges and fence posts; others have axe-like heads used for splitting lumber along the grain with a forceful overhead swing.

HACKETT SAYS

19 GET MECHANICAL ADVANTAGE FROM SIMPLE MACHINES

I want to get better at problem solving, and a useful way to get better at stuff is to learn more about the rules. The better you know the rules, the better you can play the game. The better you play, the easier it is to win. The game is rigged and we all lose, but the rules seem clear: Newtonian physics.

Physics grew from classical mechanics—physics explains and predicts how mechanics work and how forces will interact. Mechanics is based on the six simple machines: lever, wheel and axle, pulley, inclined plane, wedge, and screw. These are it—there are only six ways to manipulate force. Mix and match and stack and slice, and you have all mechanical technology. The simple machines are based in the concept of mechanical advantage.

Mechanical advantage (and all subsequent technology) sprung from Archimedes and his understanding of the lever. The genius here is twofold: First, realize that the lever, a thing everyone knows and has always understood, is not an example of something, but that everything else is an example of it. Second, if there is one rule, maybe there are others. It's not that he made the rules—it's that he was the first person to realize that there were rules.

The simple machines are more than just neat boxes to slot existing things into. They're descriptive but also predictive—once you know the basics, once you know the rules, you can combine and recombine the basic bits, and proceed rationally.

From Archimedes, we have the engineering and technology that we have now, we have the core rules that underlie everything. More directly, we have the lever—still the best way to break up pallets.

20 KNOW YOUR FASTENERS

Things fall apart. And when things do, inevitably, come to pieces, it falls to us to put them back together again. As a maker, battling inevitable entropy is just part of the fun. When attempting to bring some semblance of order to the chaos around you, knowing the right fastener for the job will help you keep it together.

DOWELS In a fastener context, dowels are short cylindrical rods, often coated with adhesive, which are used to align and secure parts together by mating with matching holes. They are most often used in woodwork.

NAILS From tiny tacks to enormous railroad spikes, nails are essentially straight, pinlike fasteners mostly made of metal. Pounded into a surface with a hammer, pneumatic gun, or any hard object in a pinch, nails hold together structures from birdhouses to barns.

RIVETS A rivet is a permanent, single-use fastener made of a soft material that's designed to be deformed to create a permanent hold. It comes with one ready-made head and a shaft that passes through holes in the parts to be joined.

SCREWS At their most basic, screws are fasteners with a head and external threading. When torque is applied to the head, a screw threads into its receiving material. Some screws are designed to cut their own threads as they go, while others are designed to mate with pre-cut threads in the receiving hole. Both kinds can be used where you want a secure joint that's easy to disassemble later.

BOLTS While not everyone agrees on the distinction between screws and bolts, bolts tend to be heavier duty than typical screws. More important, where screws can usually hold up on their own, bolts are often made to work in conjunction with threaded nuts.

NUTS Hand-fastened like a standard wing nut, crowned like an acorn nut, or wrench-tightened like a hexagonal nut, all nuts are essentially fasteners consisting of a threaded hole. They're used, often with washers, to secure bolts, machine screws, or other threaded fasteners in unthreaded holes.

WASHERS The snowshoes of the fastener world, washers are thin, flat, often round plates, usually made of metal or plastic, that go between the head of a bolt or a nut and the surface against which it is tightened. They may also be used as spacers—for instance, to add clearance for moving parts to rotate without scraping.

EYEBOLTS AND SCREW EYES Eyebolts and screw eyes are fasteners with a loop on one end, and either a screw or bolt on the other. Mostly used to hang things or attach cables to objects, the eyes themselves can be either bent or forged. If you want the eye to hold a heavy load—like your butt in a hanging chair—use forged.

21 GO THREADED OR UNTHREADED

As a general rule, threaded fasteners are more expensive and take longer to install, but offer the advantage of reversibility: You can take stuff apart again for repair or storage. Meanwhile, unthreaded fasteners may or may not be stronger, but they will almost certainly be cheaper and quicker to put in place. Threaded fasteners subject to vibration usually require additional parts—like lock washers or thread adhesives—to keep the threads from slowly working loose.

22 DECODE SCREW HEADS

Anyone who has ever put together toys or furniture with "some assembly required" knows his or her way around at least a handful of screws. But the list of fasteners is long, stretching far beyond those that can be tightened with a flat-head or a Phillips. Before you grab the wrong tool, learn to pick a head out of a lineup.

23 SINK SOME SCREWS

Screw threads, which are essentially an inclined plane wrapped around a cylinder, translate large rotational movements of the screw and driver into small, but very powerful, linear movements of the screw itself. A single turn does not advance the screw in or out of its hole very far, but that small distance is traversed with great force—enough to cut threads in woods or secure heavy loads.

The key difference between drivers is its point, shaped to match a fastener type. The boss of all drivers is the screwdriver, but even he contains multitudes: The most common are Phillips (marked with a + on its business end), flat (–), Robertson (square), and Torx (star). Regardless of your fastener flavor, drivers all pretty much work the same:

STEP 1 Establish the fastener head type and select a driver that fits. This is the hardest part—screws are simple, screw politics are complex. I recommend investing in a bit kit and an open-ended driver.

STEP 2 Hold the screwdriver in your dominant hand. Insert the tip into the recess in the head of the screw. The fit should be tight—if there's slop, you've got the wrong bit.

STEP 3 Start by pressing firmly with the screwdriver shaft along the screw axis. Once aligned, turn the driver clockwise to tighten and counter-clockwise to loosen. (It's okay to mutter "righty tighty, lefty loosey" while operating a screwdriver. Everyone does it.)

24 BEAT SPECIALTY SCREWS

Junked electronics are filled with wonderful things. I'm talking really good stuff—stepper motors, high-powered magnets in hard drives, and more. Unfortunately, getting at that good stuff requires getting through a cavalcade of exotic specialty screws. Solution: Make them all into not-so-exotic flat-head screws.

I use a rotary tool with an abrasive cutoff wheel, but a hacksaw or a gently applied 4½-inch (11.5-cm) grinder cutoff wheel can work. First, clamp the enclosure to the worktable and locate all those pesky screws—some might be hidden on the backside or under stickers. Then use your tool of choice to cut slots across the diameters of the offending screw heads. (Sometimes the factory

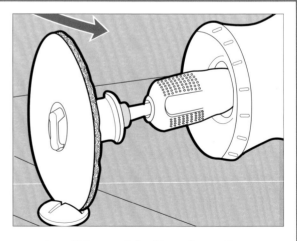

geometry will be a useful guide, and sometimes you'll have to freehand it.) Then unscrew the new, improved screws, and enjoy your bounty.

25 DEAL WITH A PESKY STRIPPED SCREW HEAD

Stripped screw heads are an annoying little problem that has befuddled do-it-yourselfers all and sundry. Luckily, there are a few techniques that can help:

A Place a thick rubber band over the head of the screw and, using a screwdriver that's a size bigger than what the screw is made for, apply hard, slow force to turn the screw.

B Use a hammer to drive your screwdriver deeper into the screw head, then rotate the driver and the screw as one.

C Glue the head of the screw to the screwdriver and twist.

D Use vise grips to grasp a thick screw head and then rotate the grips incrementally to loosen the screw.

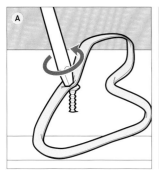
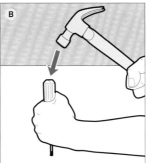
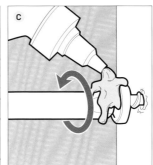
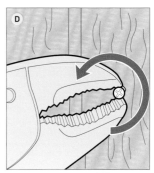

26 PUT MORE TORQUE ON A SMALL SCREW

The difficulty of turning a screw is directly proportional to its size: The bigger it is, the larger the surface area of its threads, and the greater the opportunity for friction. Today, robot manufacturing and rigorous manufacturing standards have allowed even small screws to be installed with impressive amounts of torque, especially in high-performance devices like hard drives. The small jewelers' screwdrivers that fit these screw heads aren't very wide, and their handles sometimes don't have enough girth to give you the mechanical advantage you need to get these screws loose. Here's how you add girth.

STEP 1 Set the thing you want to open on a table and insert your tiny screwdriver. The little electronics screwdrivers with ribbed barrels work best. Then take lockjaw pliers (vise grips are perfect) and grab the barrel of the screwdriver tightly.

STEP 2 Hold one hand over the butt of the screwdriver and the other on the end of the pliers' handle. Apply downward pressure by pressing your chest into the back of the hand holding the screwdriver, and let the weight of your torso keep the driver firmly in the orifice.

STEP 3 Turn the pliers, slowly and steadily, while maintaining downward body pressure. Both are needed—without the downward force, the screwdriver will take the path of least resistance, climbing out of the screw head and probably stripping it.

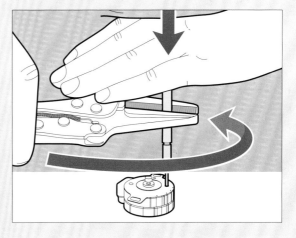

27 PICK A WRENCH

Most makers know two wrenches: a jolt of pain caused by, say, lifting a heavy toolbox, and the simple, elegant tool that, ironically, makes loosening and tightening fasteners less painful. I advise you to skip the first one (lift with your legs!) and embrace the second.

MONKEY WRENCH A classic that's typically used for tightening fasteners such as lug nuts, bolts, and screws. Not as common as a crescent wrench, but a lot more solid and stable. The distance between the jaws is changed by turning a thumb screw, usually located right below the lower jaw.

TORQUE WRENCH For precise mechanical parts with very picky torque requirements, this socket model comes with a handy display that shows just how much rotational force you've applied.

SOCKET WRENCH Efficient tool with a ratcheting function at its head that allows you to tighten or loosen a hex bolt or nut without having to re-index the wrench at every turn. The square socket shaft lets you attach different-size sockets, extensions, and accessories. Easy on the wrists.

BOX WRENCH With two enclosed rings on either end, the box wrench comes with either a six- or twelve-point recesses, both of which are ideal for moving hex-head fasteners with a very small amount of torque. Great for heavy jobs and not likely to damage fasteners.

COMBINATION WRENCH This tool boasts two different heads for two different purposes: a brute open end that helps you budge a tight nut or bolt and a closed box end for precise tightening or loosening. Both ends customarily fit the same size hardware.

SPANNER WRENCH Use on large-diameter retaining rings or fittings with holes, slots, or tabs. There are three common spanner jaw shapes: hook, C, and double hook. Often used for installing disks on angle grinders.

CRESCENT WRENCH Perhaps the most generic wrench of them all. Think of the crescent wrench as many open-ended wrenches in one, with the distance between the jaws arbitrarily adjustable via the thumb screw on its base. A proper box- or open-end wrench is a better choice, if you have one, but nothing beats a crescent wrench for versatility and portability.

PIPE WRENCH Bears a strong resemblance to the monkey wrench; the most crucial difference is that its jaws are not parallel. Apply pressure and the top jaw will shift slightly forward, biting and tightening its hold on a pipe or round stock.

28 IMPROVISE WHEN YOU HAVE NO SPANNER

Lurking somewhere in most American shops is a hook spanner, a tool with a C-shaped opening designed to grab specialty nuts that spin with axles—for example, on bikes and surface grinders. They get bought for use with a specific tool, then immediately lost. After the first time I misplaced it, I realized that while it's handy to have around, it's not necessary. Here's how to go without one:

STEP 1 Examine the shaft collar or nut you need to remove. Line up flat-head screwdrivers in the slot meant for the spanner hook until you establish the closest fit.

STEP 2 There may be a way to lock the shaft—perhaps a pit at the tip. If there's no obvious lock, secure the shaft with a vise grip.

STEP 3 Grab the screwdriver that fits, insert it into the nut, and whack the handle as hard and fast as you can with a hammer. Just once should be plenty to make the nut break loose so that you can unthread it the rest of the way by hand. Its static friction overcome, it's all yours.

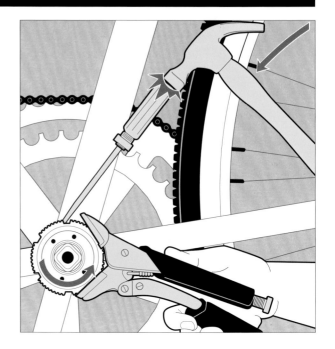

29 BOOST LEVERAGE WITH A CHEATER BAR

Even with a sturdy wrench, it may be difficult to generate enough torque to free a stubborn fitting or fastener using just your arm and shoulder. For added oomph, try adding a pipe extension, a.k.a. cheater bar, which is just a piece of pipe slid over the tool handle to increase leverage and multiply the force you can apply. A cheater bar can be mighty handy, especially when you're working on old, heavy, rusted equipment or pipes that have sat for years or decades undisturbed.

For safety's sake, make sure you slide the pipe extension down the entire length of the wrench handle. Also, keep in mind that, while longer cheater bars will give you correspondingly larger mechanical advantages, they also increase the likelihood that you will bend or break the wrench itself.

If you're working on pipe fittings, always use an additional wrench for support, especially if you're using a cheater bar. The goal is to keep from working adjacent pipe segments or fittings loose as you turn. Fasten your support wrench as close as you can to your action wrench, and hold the one firmly in place while you turn the other.

30 WIELD A COMBINATION WRENCH

Hang around a shop for five minutes, and you're bound to spot a combination wrench. With an open-ended head on one side and a box head on the other, it's the go-to tool for cracking big bolts and nuts loose from slabs of heavy machinery. Basically, you use the open-end wrench to get them started, then switch to the box-end side to spin them out.

STEP 1 Figure out the size of the wrench you need. Be careful (but not too careful—hesitation smells like weakness!). **Pro tip:** Try a 9/16-inch (14.25-mm) wrench. Chances are, you'll be right. If not, laugh it off.

STEP 2 Slide the wrench's open end across two parallel faces of the fastener. Hold the jaws in place on the fastener head with your nondominant hand, pressing down—the wrench's body should pop up out of the plane. To loosen, grab the free end and push it away from you in a counterclockwise motion. If it won't budge, check for rust or grime and spray with solvent or rust breaker, wait, and try again. If you still need torque, reverse the wrench so that you're pulling toward your body and try planting a foot on a sturdy object. If it still won't budge, consider that reverse-threaded fasteners do exist and, though they are rare, you may have encountered one. If you're sure you've been turning the right way, reach for the cheater bar.

STEP 3 Once the first half turn is completed, do another, and it should be loose enough to switch over to the 12-point head. Keep the 12-point on the fastener and spin the handle around and around until the fastener is free.

STEP 4 Oh no—does it just spin and spin? Sometimes fasteners go all the way through an object and are held in place with a backing nut. These nuts can be hard to get to and difficult to hold a wrench on. I like to put a 12-point or a vise grip on the backing nut, and let it spin until it jams itself—the part that was getting in your way will now serve you, holding the wrench where you need it. Backing nuts are often one increment (usually 1/16 inch [1.5 mm]) wider than the hardware head. So to back a 9/16-inch (14.25-mm) bolt, you need a 5/8-inch (16-mm) wrench.

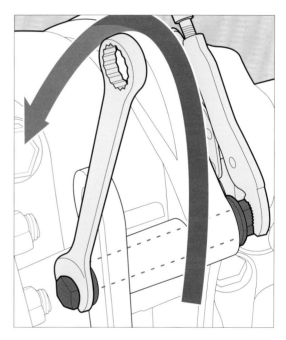

"Wrenches are relatively new on the maker scene—mainly because they're needed for threaded fasteners, which didn't really exist until the screw-turning lathe was invented in the Industrial Revolution. Now you know."

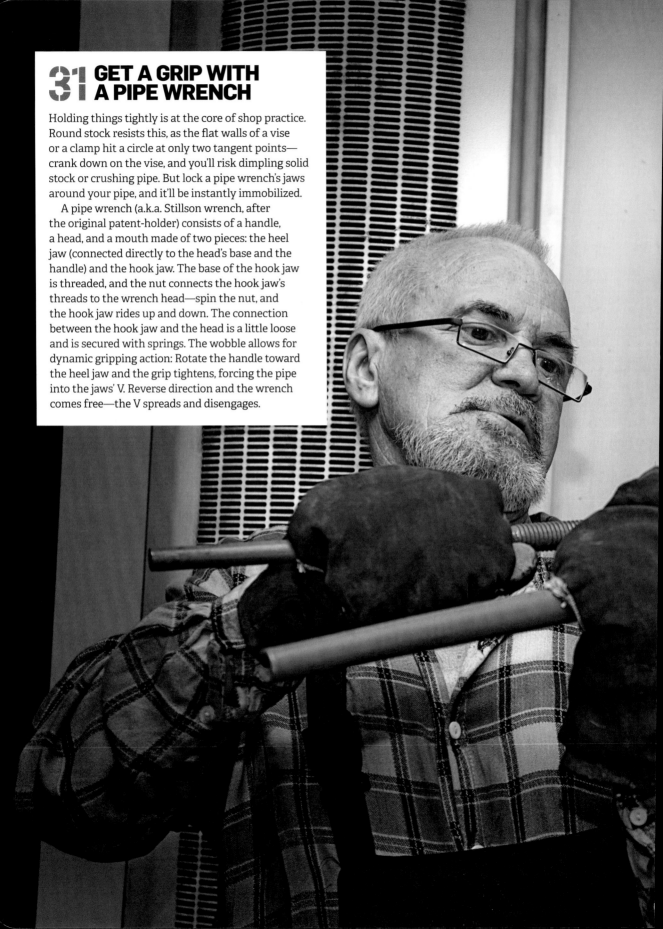

31 GET A GRIP WITH A PIPE WRENCH

Holding things tightly is at the core of shop practice. Round stock resists this, as the flat walls of a vise or a clamp hit a circle at only two tangent points—crank down on the vise, and you'll risk dimpling solid stock or crushing pipe. But lock a pipe wrench's jaws around your pipe, and it'll be instantly immobilized.

A pipe wrench (a.k.a. Stillson wrench, after the original patent-holder) consists of a handle, a head, and a mouth made of two pieces: the heel jaw (connected directly to the head's base and the handle) and the hook jaw. The base of the hook jaw is threaded, and the nut connects the hook jaw's threads to the wrench head—spin the nut, and the hook jaw rides up and down. The connection between the hook jaw and the head is a little loose and is secured with springs. The wobble allows for dynamic gripping action: Rotate the handle toward the heel jaw and the grip tightens, forcing the pipe into the jaws' V. Reverse direction and the wrench comes free—the V spreads and disengages.

32
HACK YOUR PIPE WRENCH

The Stillson wrench is the single best shop tool for gripping anything that's round—stock, dies, jammed Jacobs chucks, or extra-wide reams. More often than not, when I'm using a pipe wrench, I'm not actually using it on pipe. Here are some other ways to use this go-to tool:

USE IT AS A SUPER VISE If you want to hold round work still while you really go at it with another tool, grab it with a pipe wrench and then clamp the wrench handle into a vise. Make sure your shop motions (grinding, milling, drilling) are being done with the correct orientation: The tool should move toward the heel jaw to keep the object lodged in the pipe wrench's counterclockwise grip.

PROTECT AGAINST FIERCE GRIP Pipe wrench jaws are hardened steel, the better to grip and bite into pipe. They will mark anything softer than hardened steel and can be damaged by anything harder. Blunt the edges by wrapping your work with a rag or handkerchiefs.

33 PICK A SAW

Saws are essentially tools used to cut hard materials like plastics, metals, and wood. Little other than a cutting mechanism and a handle (and sometimes not even that!), handsaws are about as varied in type and size as the jobs for which they're used.

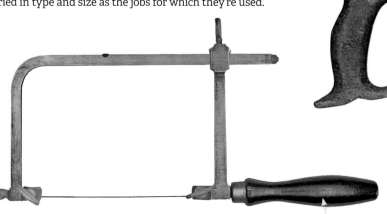

CROSSCUT SAW Stepsibling to the ripsaw. Knifelike teeth make it ideal for cutting wood against the grain. Some have teeth that alternate cutting and pushing sawdust, and they often cut on both the push and the pull stroke.

HAND JIGSAW/JEWELERS' SAW Your best friend when it comes to cutting out small and intricate shapes in soft metals, wood, fiberglass, or similar materials. Though blades often break, lubrication can help minimize the risk.

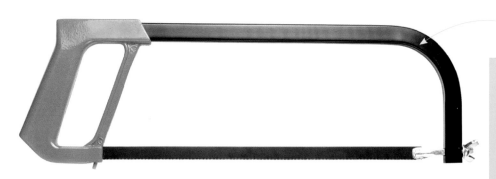

HACKSAW Best for cutting plastics and metal pipes. Can cut a variety of materials, depending on the blade. The heavier or thicker the material, the coarser and larger the teeth on your blade should be.

COPING SAW Used for, well, coping—cutting imperfect corner pieces (or pieces to fit an imperfect corner) so that they create a clean joint. U-shaped frame and blade on a swivel allow for easy interior and turning cuts.

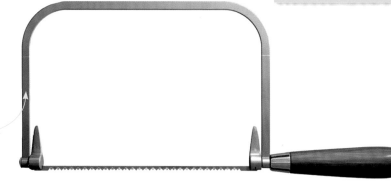

RIPSAW Primarily used to cut wood in the same direction as the grain. Its teeth are almost chisel-like, allowing the ripsaw to cut straight lines regardless of the grain. Many only cut on the push stroke.

WIRE SAW Admittedly not the ideal choice for everyday sawing, but absolutely the best saw that fits in your pocket. Essentially braided steel, and sometimes equipped with teeth. Works best to cut wood, plastic, and bone, particularly if you're stranded in the woods.

SHEETROCK SAW Similar in shape and purpose to the keyhole saw. Small size and coarse teeth make it ideal for cutting things like holes for electrical outlets in drywall, often without even drilling a starter hole first.

KEYHOLE SAW Originally intended to cut keyholes. Small stature makes it great for cutting in places a larger saw might not fit. Often used to cut softwoods or drywall, its abilities change with the blade you use.

JAPANESE FLUSH-CUTTING SAW Makes cuts flush with a surface. Flexible blade that cuts only on the pull stroke allows it to hug tight to a surface without scratching, often making more precise cuts than many power tools.

34 FOCUS ON WOOD

Wood is the fibrous structural component of trees—the part that holds the tree up and provides the support system for the bits that feed it, water it, and make more trees. Taking a cue from the original owner, we use wood as structural support, and also as a shapeable, somewhat plastic medium to bend or carve to our desires. Basically, if something can be made, it can be made out of wood. Here are some things to know about it.

DIMENSIONAL LUMBERS HAVE NAMES THAT ARE NUMBERS, which should be thought of as just a name and not a descriptive measurement. In general, subtract ½ inch (1.25 cm) from each number to get the true size, ¼ inch (6.5 mm) if the first number is "1." For example, a 2x4 is actually 1½-by-3½ inches (3.75-by-9 cm). Essentially, 2x4s are required to measure 2-by-4 feet (0.6-by-1.25 m) when they leave the mill, but they often shrink during the drying process or when they're milled further to remove defects.

DROPS MAKE GREAT JIGS AND BRACES.

NEVER BURN PRESSURE-TREATED LUMBER.

DON'T KNOW IF YOU'RE DEALING WITH A HARD- OR SOFTWOOD? SMELL IT. THERE ARE A NUMBER OF DIFFERENT WAYS TO ID WOOD (LIKE SANDING TO EXPOSE THE GRAIN), BUT MY METHOD IS TO SHAVE OFF SOME CLEAN WOOD AND SMELL IT. SOFTWOODS WILL SMELL LIKE PINE, OAK WILL SMELL LIKE AUTUMN, CHERRY SMELLS LIKE MONEY.

WOOD IS UNFORGIVING—I NEVER WORK WITH IT. IT SCARES ME. CUT A GOUGE IN A THREE-FOURTH FINISHED PIECE OF WOODWORK AND, IF YOU'RE LUCKY, YOU CAN REMOVE THE WHOLE BOARD AND PUT IN A NEW ONE—NOT SO MUCH IF IT'S ALREADY GLUED IN. PUTTY LOOKS HORRIBLE, AND THERE'S NO WELDING ON WOOD. SO MEASURE PRECISELY, GO SLOW, AND BE GINGER WHEN FITTING JOINTS TOGETHER.

IMPORT FROM THE TROPICS.
Here, tropical hardwood is a valuable commodity, but in Southeast Asia it's the most convenient wood for pallets. Find places near you that import stuff directly from that part of the world; there will be stacks of pallets that you can pull into individual boards with a claw hammer. Don't speak the language? Bring beer—beer speaks the language of guys who work on loading docks worldwide.

ENGINEERED WOOD IS WOOD PLUS SCIENCE, THE HOTDOGS OF THE WOOD WORLD. It's designed to solve tree limitations by stacking the structure of wood fibers in a more rational way, or using glue to turn sawdust into a useful material. Not pretty, not natural, but useful in every aspect of making. Build your worktables from plywood or medium-density fiberboard (MDF), and when the surface gets gouged or burnt, just throw a new layer on top and screw it to the old. Repeat as needed.

PRESSURE-TREATED LUMBER GETS ITS GREEN COLOR IN A PESTICIDE PRESSURE COOKER. USE IT FOR ANY STRUCTURES THAT ARE UNDERGROUND (LIKE TUNNELS OR SUPPORT POSTS).

NATURAL WOOD IS AS IT COMES FROM THE TREE, REFLECTING THE TREE'S CHARACTERISTICS. THE TREE'S YEARLY GROWTH IS EXPRESSED IN RINGS THAT BECOME THE **GRAIN** OF THE FINISHED LUMBER. POINTS WHERE BRANCHES JOINED THE MAIN BODY OF THE TREE ARE VISIBLE AS **KNOTS**, AREAS THAT ARE USUALLY FAR MORE DENSE AND BRITTLE. WOOD COMES FROM INDIVIDUAL PLANTS, BUT TREE FARMING, INDUSTRIAL LUMBER MILLING, AND MILLENNIA OF PRACTICE YIELD PRETTY CONSISTENT LUMBER. WHEN A PIECE OF WOOD HAS UNIQUE FEATURES, LIKE SPRAWLING KNOTS OR **BURLS** (BASICALLY TREE TUMORS), IT'S CONSIDERED A FEATURE AND HIGHLIGHTED.

NATURAL WOOD IS USUALLY DIVIDED INTO TWO CATEGORIES: HARDWOOD AND SOFTWOOD. Technically, hardwood is from deciduous trees, those with broad leaves that shed in the fall, such as oak, maple, cherry, and mahogany. It's also expensive and usually has a look-at-me quality. Softwood is from coniferous trees, those with needles or denser, thicker leaves that stay on year-round, like pine, cedar, and fir. For maker purposes, hardwoods tend to be denser, tougher, and slower to grow than softwoods, making them more durable, with a finer texture. Fast-growing softwoods are less dense, less tough, and have lower structural strength, but they grow straight and quickly, making them our primary source of lumber. Use them in low-stress places where you need wood to work, not get looked at.

BUY THE STRAIGHTEST BOARDS.

35 FAMILIARIZE YOURSELF WITH FINISHES

A wood finish can provide protection for your chosen aesthetic, making the grain pop or adding luster. But the flood of finishing options can be confusing to the first-time maker. There are waxes, oils, varnishes, shellacs, lacquers, and water-based finishes, all of which fall into one of two categories: evaporative or reactive. In the evaporative camp (such as lacquer and shellac), the finish dries as the solvent dissolves—these finishes can be stripped away by reapplying the evaporative component. Meanwhile, reactive finishes (such as linseed or tung oils and varnishes) are locked into the wood via a chemical reaction, so they tend to be more durable.

TO APPLY AN OIL FINISH

STEP 1 Make sure you have a clean work area so a gust of wind won't blow dirt onto your wet piece; several fresh cotton cloths to apply the finish; the ability to access your piece from all sides; and sufficient light to see it from different angles.

STEP 2 Sand the wood with the grain up to 180 or 220 grit, then clean off and remove the resulting sawdust—a tack cloth works well.

STEP 3 Next, flood the area with a 1 part pure tung oil to 1 part mineral spirits or citrus solvent mixture. Use a smooth cloth to push the pool around your piece, allowing the oil to seep into the pores of the wood. Repeat until you have evenly covered your project with oil.

STEP 4 Let the wood absorb the oil for 30 minutes, then remove any excess with a clean cloth. Give it 24 hours to cure. (Wood grain can raise up after the first few coats of finish dry. If needed, give it a light sanding to remove raised grain.)

STEP 5 Reapply oil to your piece three to four more times, with 12 to 24 hours of time in between each application. At the end, rub on a layer of wax. **Note:** To avoid spontaneous combustion from the oil-soaked rags reacting with oxygen, spread them out to dry before discarding.

36 MASTER BUTT JOINTS

Some things get stupid names because they themselves are just so basic, so commonsense that giving them a real name endows them with undeserved dignity. The butt joint may be one of those things. It comes in two main types: the mitered butt joint and the plain old butt joint. For the plain version, you simply affix the ends of two boards together with wood glue, then reinforce with nails or screws. For the mitered variety—which is only better because it's prettier—follow these steps:

STEP 1 To join two pieces of wood at a 90-degree angle, you'll first need to mark each piece of wood with a line at a 45-degree angle, drawn from front to back at the corner of each piece. Mark this 45-degree angle on the opposite sides of each piece, and join the bottom corners of those lines together with a straight line on the third side. This will mark out the triangular wedge of wood you'll be removing from each piece.

STEP 2 Start your cut at one of the 45-degree lines, cutting a shallow groove. Finish your cut carefully, checking that it lines up with both the end corner of the wood and the straight line that you first drew. Repeat on the second piece of wood.

STEP 3 Once you've cut each end into a 45-degree angle, line the pieces up with the newly cut surfaces touching. Check to make sure this forms a 90-degree angle, then glue the ends together with wood glue to form a mitered butt joint. Clamp the joint while it dries. Nail or staple in place.

37 SAND IT RIGHT

History points to 13th-century China as the first culture known to use sandpaper—at one time, they even used dried sharkskin. After you've made something from new wood, the final stages include sanding and finishing. Primarily, you sand wood to remove machine marks and create a smooth feel ready to accept a finish (such as paint, wax, or oil).

Today a variety of sanding papers and devices are available. Grit rating—e.g., 36, 80, 120—refers to the size of the particles attached to the sandpaper, sanding disc, or sanding belt, measured in microns. If you look closely at a single sandpaper particle, you will see facets that abrade the wood fibers. The lower the grit number, the larger the particle size—thus the larger, more aggressive individual cut marks in the wood. Much higher grit numbers are used to polish.

In general, you sand with a lower grit number and then move to the next higher grit number. For instance, first sand with 100 grit to remove saw marks. Then move up to 150 grit to remove scratches made by the 100 grit, making the wood softer.

Try sanding first with just your hand backing the sandpaper, then with a flat block backing the sandpaper so you can move in the same plane as the wood, and then with a random-orbit sander to experience the joy of not tiring yourself out.

38 SHEAR AND SAW

Making is all about cutting things apart and putting them back together in a certain order. As a category, cutting is broad and seemingly without bottom—after all, this is a category that houses stone wedges and lasers, ultrasonic sound and pocketknives. With all of its complexity and diversity, cutting can still be split into two distinct parts: shearing and sawing.

SHEARING Shearing cuts split material along existing fault lines (e.g., a wood-splitting maul), or uses the mechanical advantage of a wedge to transfer considerable force to thin cutting edges (e.g., knife, scissors, or tube cutter). When the force at the tip of the wedge is higher than the shear strength of the material, the material is cut. While nothing is removed or lost, shear force can distort and tweak the sheared material.

SAWING Sawing covers all cutters that remove material: saws, of course, but also drills, laser cutters, and alien saliva. Most are doing the same thing as shears, but in smaller blobs—sawdust and metal chips are basically tiny little sheared-off bits. The material lost in the process is known as the kerf, a neat word that will make people think you know stuff. When planning builds, account for kerf—the width of the saw blade—or you will fall short.

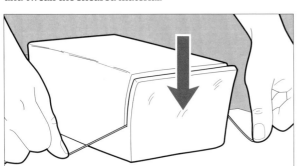
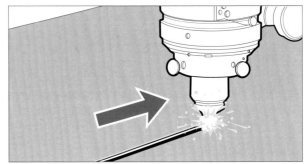

39 USE A TUBE CUTTER

Tube cutters are made to do one thing, and one thing only: Cleanly cut nonferrous tubing. I used to think they were a solution without a problem, but once you've seen a slanty, slow, burr-covered hacksaw cut, you will see the light. Copper, PVC, or plastic—as long as it is rigid and not as strong as steel, tube cutters will work fine.

STEP 1 Use a sharp pencil or scribe to mark where you want the cut. The actual cutting wheel has a very narrow profile—about the same as a box cutter blade—so marking with a marker or piece of chalk gives too much room for error. **Pro tip:** A box cutter will give a nice, crisp marking line on most nonferrous materials.

STEP 2 Loosen the set screw until the pipe can fit between the rollers and the cutting wheel. Tighten the set screw until the cutting wheel just divots the tube. Grab the set screw handle and spin the cutter around the pipe a couple of times. Tighten, repeat, and after a few turns you will feel something give. Keep spinning another turn or two and the tubes will separate.

STEP 3 Time to deburr, or get rid of all the sharp jagged crumbs. The side of the tube cutter has a deburring tool—a crude, sharpened triangle. Jam the tip into the tube, rotate while applying pressure, and interior burrs will be shaved away.

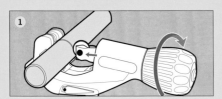

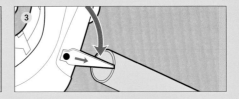

40 CHOOSE SHEET METAL SNIPS

Sheet metal snips take on metal as scissors do paper. Use them to cut and trim auto bodywork, HVAC duct, radiator grating, and stovepipe—anything sheet metal. Like anything else, snips come in a variety.

STYLE Snips come in two styles: traditional, or tin snips, and compound, or aviation snips. Tin snips use the mechanical advantage of their long, straight handles to cut, while aviation snips use a compound lever to get high force out of short handles and short bites. Tin snips' long lever requires two-hand operation and results in a decrease in control; aviation snips' short cut is more than made up for by quick, one-hand operation.

COLOR More than just a pretty color, aviation snip grips let you know when to use which cutter. Red grips can cut straight or sharp left turns, and straight lines will tend metal to the left. Green grips cut straight or sharp right turns, with metal tending to the right. Yellow grips cut straight, tending metal up and down. **Pro tip:** Get a pair of pipe and duct snips. They have yellow handles and three blades. This combination of saws and shears cuts in two places at a time, giving a strip of freedom to go where you will.

41 KNOW YOUR CHISELS

Chisels have been around for a very long time. The purpose of some of the earliest known cutting tools is unclear; without context, all we see is shaped rock and attitude, which sounds about right for chisels—essentially knives that really mean it. Apply energy to the butt, either through percussive force (e.g., whacking with a hammer or mallet) or a steady push, and the blade tip will get to cutting.

A WOOD CHISELS Chisels are a primary woodworking tool, used to make any number of cuts, from giant mortise and tenons, to the tiniest sculptural detail in muscles and tendons. **Pro tip:** Chisels need to be kept scarily sharp, otherwise they will gouge rather than cut wood.

B METAL CHISELS Before the dawn of the angle grinder, metal chisels got quite a workout, less so now. Metalworking chisels fall into two categories: hot chisels (used for cutting, slotting, punching holes, and forcing shapes into hot metal still plastic from the forge) and cold chisels—the pre-grinder go-to for freestyle metal cutting and trimming solid stock.

C MASONRY CHISELS Chisels are still widely used in masonry. While some are used for carving, they're mostly a tool of destruction. Jackhammers, powered by compressed air, are essentially heavy-duty chisels.

42 PICK A VISE, CLAMP, OR JIG

Sticking things together is the core of making—everything that comes before is prep, and everything that comes after is finishing. Having clamps is like having an extra, stronger, more resilient set of hands to hold things in place while you work.

ANGLE VISE Allows you to secure a piece and tilt it to a specific angle so you don't have to shift the head of the drill press or milling machine.

CORNER CLAMP Holds pieces at exact 90-degree angles to one another, as if laying out a picture frame. When the cut of a piece is not quite perfect, this clamp crushes, bends, and forces the ill-cut bits into regimented order.

TAPE From electrical to duct, masking, and gaffer tape, the list of sticky-material strips is long (see #73). Use to seal low-pressure holes, join materials temporarily, or as a sign of shoddy craftsmanship.

GLUE There are a million kinds for a billion uses, but all adhesives connect material faces to one another by creating a bridge material from one set of surface imperfections to another.

PIPE CLAMP Made up of two gripping surfaces, a hand wheel to tighten/loosen, and whatever length of standard plumbing pipe. Use to jig up that trailer or carnival ride and hold it all in place as you work.

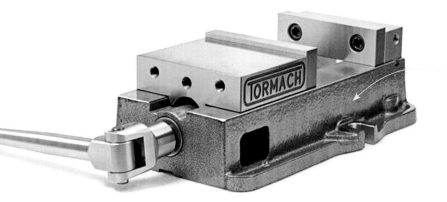

MACHINIST VISE Heavier duty than a bench vise. Used to hold things securely so a machine tool can have its way with them. The angle vise—a subset—can be adjusted to tilt and pan to different angles.

C CLAMP Operates like a free-floating (or sometimes stationary) bench vise. Used to hold things together securely by leveraging the power of the inclined plane. Beware with softer materials, as the gripping edges focus force into a small point and can damage surfaces.

PIPE HANGER/ZIP TIES/ BALING WIRE/HANGER WIRE All are more flexible than clamps and extremely useful for mocking things up—so useful, in fact, that sometimes you may just leave the temporary fix forever.

BENCH VISE Holds stuff still while you sand, saw, or otherwise manipulate it. Comes in different sizes of widely varying quality. Works better than gravity does when you're waiting for glue to dry.

HACKETT SAYS

43 USE AN OLD-SCHOOL TOOL FOR A MANUFACTURED EFFECT

I did not grow up around tools and making. Other than an unhealthy obsession with knives, my curiosity about how things worked involved carefully taking things apart, attempting to put them back together, eventually losing interest, and then getting in trouble for ruining a perfectly good clock radio. I knew which end of a hammer to hold, and I knew what drills did, but that was as far as things went.

On the rare occasion I was called upon to assemble some furniture or fix something, I would quietly panic. Coupled with decent strength and a willingness to blaze through, I left a trail of destruction in my wake. Split wood, stripped screws—the only thing stronger than my incompetence were threaded fasteners. There was a right way and a wrong way, as with anything, but this one had a catchphrase: "Righty tighty, lefty loosey." I could be crap at everything else, but nuts and bolts were idiotproof.

In fact, nuts and bolts are a shorthand for the best parts of our civilization. Exact, interchangeable, consistently produced—even the crap ones were better than what came before them, and they work even when guided by an idiot.

Time passed. I got interested, then obsessed, with how stuff works and making. I got less bad, then okay, then pretty good, but things clicked when I realized that, of all the references to taps that I encountered, none of them was about beer.

Drill the right hole, run a tap, and every bolt ever made in that size will fit. A couple of minutes' work and your creation is seamlessly, tightly joined to the best our civilization has to offer.

44 TAP A HOLE LIKE A PRO

The advent of the screw-cutting lathe and standardized threaded fasteners made the Industrial Revolution possible. Ever wonder how they made the fasteners for that lathe? They used a tap—a handy insert that, when twisted through metal, will create internal threads, transforming mere holes into nuts. The flip side of that coin is the die, which cuts external threads, turning rods into bolts. For short runs, repairs, or anyone who doesn't have a lathe, taps and dies are a simple, straightforward way to make raw materials into professional-looking (and -acting) parts, integrate the DIY seamlessly, and repair and customize things to your specifications.

Taps and dies are sized in the same manner as threaded fasteners: the major diameter, followed by the number of threads per inch. A very common size is ¼ – 20. It indicates that the piece is ¼ inch (6.5 mm) in diameter at its widest point, with 20 threads per inch. Each tap is paired with a drill bit that will allow for threads of a specific depth.

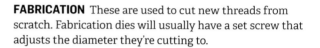

FABRICATION These are used to cut new threads from scratch. Fabrication dies will usually have a set screw that adjusts the diameter they're cutting to.

REPAIR Repair taps and dies are used to fix threads that are damaged. One pass will recut threads.

PLUMBING Threaded plumbing connections are tapered to allow for a fit that tightens as it turns.

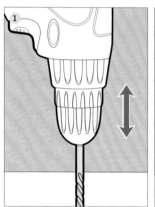 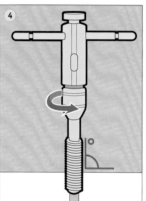 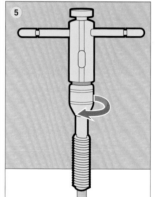 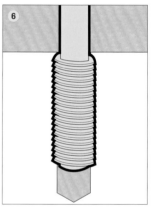

TO TAP

STEP 1 First, you need a hole. Consult a tap chart and drill the hole you need. If it's a blind hole (a hole that you can't see the bottom of), drill it at least 1/8 inch (3.25 mm) deeper than the depth of thread will be.

STEP 2 Clear the top and bottom of the hole so it's free of burrs. Chamfer to create a beveled edge along the top, if you are feeling fancy.

STEP 3 Add a drop of metal-specific or pipe-tapping fluid to your hole. **Pro tip:** In a pinch, kerosene makes a great tapping fluid for aluminum, bacon grease for steel.

STEP 4 The start of the cut is the most important part. To make sure the tap is square in the hole, slide a square or known true block of scrap under the tap handle on all sides, checking front to back and right to left. Turn the tap handle clockwise, slowly and steadily, until you feel it begin to bite the metal. Check one last time for trueness.

STEP 5 Time to tap. Seize the handle and turn it clockwise one full turn. At the end of every turn, back the tap one half turn counterclockwise—you'll feel chips popping free. Keep going, slow and steady, until the tap pops out the other side of the hole. If the tap gets reluctant, don't force it. Breaking a tap inside a hole will ruin your day. Back out all the way, clean the tap, re-lube, and try again.

STEP 6 Once the hole is threaded, spin the tap counterclockwise until it rotates out of the hole. Grip it firmly all the way out, paying special attention to the last turn or two—this is when attention wanders and holes are messed up by the tap falling over, ripping threads or cross-threading.

45 THREAD A PIPE WITH A DIE

Dies aren't used as much as their tap brothers, but they'll still help you thread a bare pipe when you're making custom builds and a standard, threaded pipe just won't do. (Which, when you're making custom builds, is pretty much always the case.) Also, bare pipe is significantly cheaper than threaded.

TO THREAD A PIPE

STEP 1 Cut, taper, and clean the pipe. Cut it to length using a grinder or saw. Plumbing pipe is tapered at the connection, making for a tighter, less-likely-to-leak connection. The taper is a subtle 1 inch (2.5 cm) over 16 inches (40.75 cm), so use a grinder or belt sander to make a barely noticeable taper; use off-the-shelf pipe for guidance on creating the angle. Remove any burrs.
Pro tip: Don't use the pipe cutter on steel. It seems like the correct tool for the job, but you will destroy the blade, and likely probably torque up the frame, too.

STEP 2 Clamp a pipe wrench into a vise. Position the wrench so that the upper jaw is closer to you and the wrench opens to your left. Place the pipe in the wrench with 4 inches (10 cm) of the end to be threaded protruding upward, turn the pipe clockwise to snug it in the wrench, and make sure it's square to the vise and wrench.

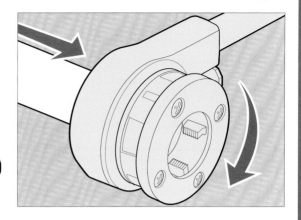

STEP 3 DIE! Plumbing dies ensure the tool is perpendicular to the pipe by sliding the pipe through the back of the die. The first couple of die teeth should slide over the taper. Set the die, then lube it generously with pipe-cutting lube, a horrible-smelling sludge that'll never completely wash out of your clothing or off your hands.

STEP 4 Turn the die until it catches, then thread the pipe one turn at a time, backing off often. It might feel a little uneven depending on your tapering job. Keep going until the first threaded bit emerges from the die, then reverse and spin the die free. Wipe the threaded end clean (it'll be sharp, so be careful of your fingers.) Repeat as needed.

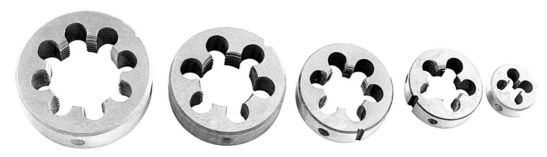

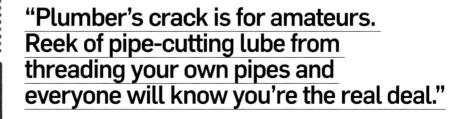

"Plumber's crack is for amateurs. Reek of pipe-cutting lube from threading your own pipes and everyone will know you're the real deal."

46 SET UP A GAS WELDING STATION

With welding, you take a few items (metal pieces, a filler metal of the same type, and an implement like a rod, wire, or stick), and come out the other side with one. The crucial condition? Heat, which is made in two ways: flame, as in oxyacetylene (or gas) welding, and electrical short circuit (covered in entries #105–110). And where there's heat, certain protections need to be in place.

SPARKER LIGHTER To ignite the gas, you need a spark. A handheld mechanical sparker uses flint to create flame without pesky, highly flammable fuel.

SAFETY GOGGLES AND GLOVES These should need no explanation, but just to be clear: Cover your eyes and cover your hands when dealing with this stuff.

REGULATORS Often available as a set, these valves safely control the flow of gas from the tank into a hose by cutting it off at a certain pressure. Oxygen needs a high-pressure regulator, acetylene low pressure.

GAS TANKS Gas welding requires a controlled flow of oxygen and acetylene, tanks of which you can purchase or rent from a gas supplier. High-pressure gas cylinders are nearly indestructible—as long as the cap is on. Once it's off, the delicate valve is exposed to damage, and the cylinder may shoot through the wall or leave holes in people and equipment. Keep tanks chained up instead.

HEAVY-DUTY FLOOR Cement is best, tile is good, and wood is totally usable as long as you don't mind the high probability of standing on a ruined floor. Stay away from carpets, rugs, linoleum, or vinyl.

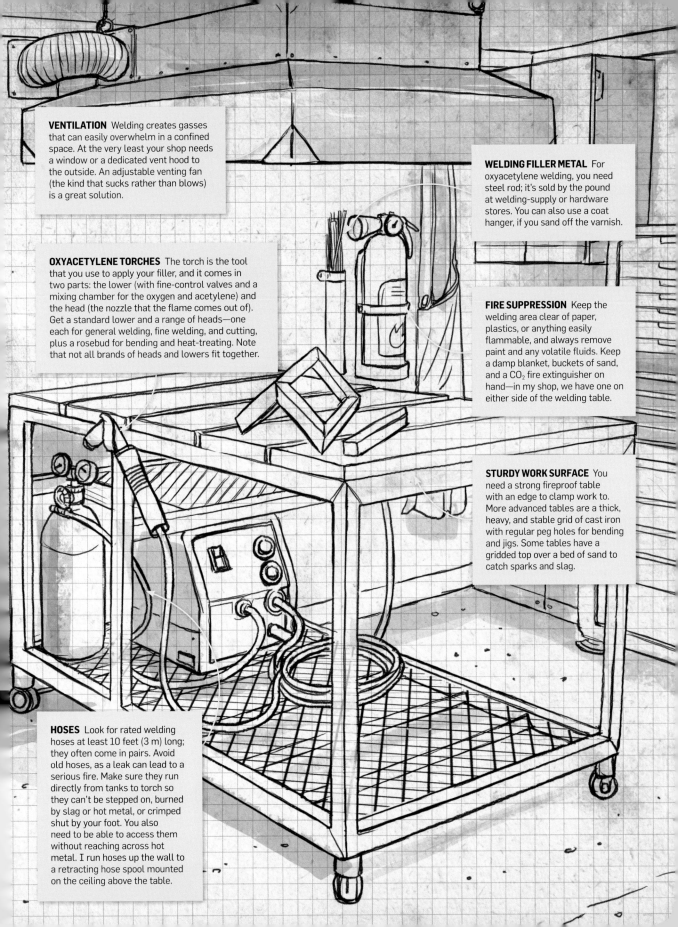

VENTILATION Welding creates gasses that can easily overwhelm in a confined space. At the very least your shop needs a window or a dedicated vent hood to the outside. An adjustable venting fan (the kind that sucks rather than blows) is a great solution.

WELDING FILLER METAL For oxyacetylene welding, you need steel rod; it's sold by the pound at welding-supply or hardware stores. You can also use a coat hanger, if you sand off the varnish.

OXYACETYLENE TORCHES The torch is the tool that you use to apply your filler, and it comes in two parts: the lower (with fine-control valves and a mixing chamber for the oxygen and acetylene) and the head (the nozzle that the flame comes out of). Get a standard lower and a range of heads—one each for general welding, fine welding, and cutting, plus a rosebud for bending and heat-treating. Note that not all brands of heads and lowers fit together.

FIRE SUPPRESSION Keep the welding area clear of paper, plastics, or anything easily flammable, and always remove paint and any volatile fluids. Keep a damp blanket, buckets of sand, and a CO_2 fire extinguisher on hand—in my shop, we have one on either side of the welding table.

STURDY WORK SURFACE You need a strong fireproof table with an edge to clamp work to. More advanced tables are a thick, heavy, and stable grid of cast iron with regular peg holes for bending and jigs. Some tables have a gridded top over a bed of sand to catch sparks and slag.

HOSES Look for rated welding hoses at least 10 feet (3 m) long; they often come in pairs. Avoid old hoses, as a leak can lead to a serious fire. Make sure they run directly from tanks to torch so they can't be stepped on, burned by slag or hot metal, or crimped shut by your foot. You also need to be able to access them without reaching across hot metal. I run hoses up the wall to a retracting hose spool mounted on the ceiling above the table.

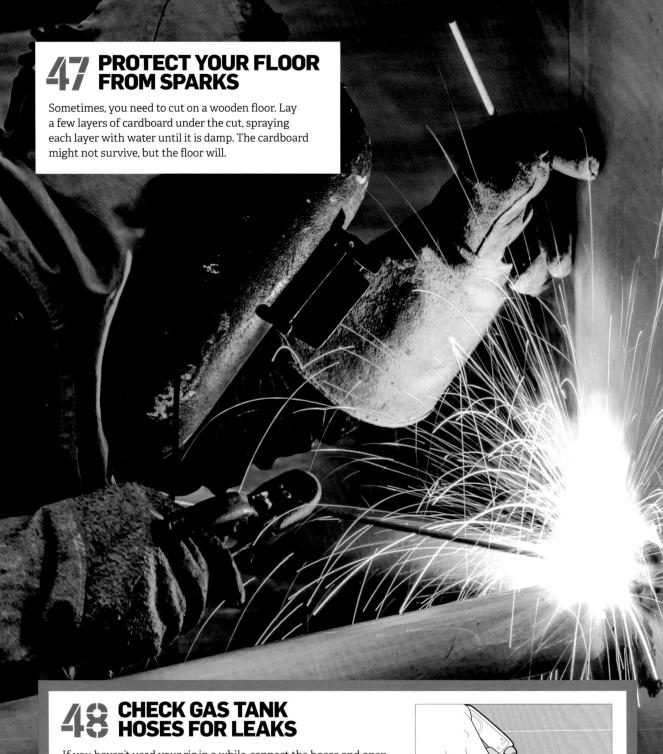

47 PROTECT YOUR FLOOR FROM SPARKS

Sometimes, you need to cut on a wooden floor. Lay a few layers of cardboard under the cut, spraying each layer with water until it is damp. The cardboard might not survive, but the floor will.

48 CHECK GAS TANK HOSES FOR LEAKS

If you haven't used your rig in a while, connect the hoses and open the regulators, but keep the torch closed. Take a spray bottle filled with soapy water and mist the length of the hoses. If you see any bubbling, that means there's a pinhole or crack in the hose. Electrical tape works for a temporary repair, but leaks should be cut out and the hose spliced into barbed connectors for long-term use.

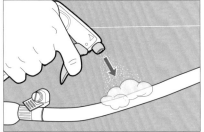

49 MAKE YOUR FIRST OXYACETYLENE WELD

Welding is the darkest of maker arts, surrounded by an intimidating, rusty wall of mystery, danger, and cool. Oxyacetylene welding is welding without power—just you, gas, metal, and flame. Once you've got a safe and functional work space set up, try your hand at making a simple joint.

STEP 1 Take the metal pieces that you'd like to join and clamp them together in the desired position and angle as tightly as possible.

STEP 2 Get down to naked metal with a grinder, sander, or file. Strip off paint or rust where it's going to be welded and at least 1 inch (2.5 cm) on either side.

STEP 3 Get your body right. Make sure you can reach and have good control at the beginning, middle, and end of the weld. Practice the movements to assess.

STEP 4 Stack filler rod within arm's reach, then make sure the regulator on the tank is open. Hold the torch and turn on the acetylene—hear the hiss from the red hose and squeeze the sparker. It'll create a feathery flame and put out black smoke.

STEP 5 Turn on the oxygen and creep it up to about 7 to 10 pounds per square inch (0.5–0.7 kg-force/sq cm). You'll see the flame tip get more focused and brighter, and the smoke will clear.

STEP 6 Position the torch near where you want your weld to start, hovering ¼ to ½ inch (0.65–1.25 cm) above it, with the tip of the flame pointing toward the junction of the metal pieces.

STEP 7 Slowly move the torch in circles 2¼ inches (5.75 cm) in diameter. The metal will glow red, then yellow, then paler and paler and drop into a reflective liquid. That's your puddle—the source of mingled, hot metal that you'll use as glue for your weld.

STEP 8 With your other hand, grab a filler rod and dip it into the edge of the puddle. After making more small circles with the torch, dot the rod at the end of the weld to secure the two pieces along the joint's length—this is called a tack. For longer lengths and thinner materials, go to the middle and do a tack, then make another tack in the middle of that tack and your first puddle, splitting the difference until you have a tack every 2 to 3 inches (5–7.5 cm).

STEP 9 Go back to your first weld and dip the filler rod to create another puddle, this time overlapping the first puddle about 50 percent. I call this, "Make a puddle, move a puddle." Do this all the way down until you've covered the joint.

STEP 10 Let it cool. Hit it with a wire brush. Take a look at the back to ensure you've made a deep weld.

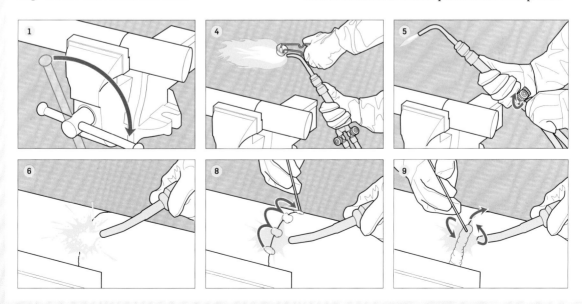

50 BRAZE WITH A TORCH

Brazing is the duct tape of welding—the ultimate adhesive, securely affixing anything to anything, like steel to aluminum, cast iron to copper. Only it's not welding: The base metals don't mix. Instead, they're joined at the microscopic level like a solder joint, with brass or bronze flowing in as the glue. Here's how:

STEP 1 Prepare a fairly close connection, leaving a little space for the brazing rod to flow. Clean the joint well—get down to the shiny metal or the braze will not stick.

STEP 2 Mix flux with water and use a paintbrush to get it into every nook of the joint. Flux absorbs oxides, keeping the joint pure as it heats up. General-purpose fluxes are fine, but borax and water work, too.

STEP 3 Assemble the joint, snug but not tight. Use baling wire as a spacer, jig, or lashing for simple or complex joints. It can bend to any shape and take the heat, and it is endlessly reconfigurable.

STEP 4 Move the torch around to heat the joint evenly, spending more time on thick areas. Watch the flux—it will sizzle off the water, get clumpy, then liquefy and turn clear. When the flux is clear everywhere, the metal is hot enough to braze.

STEP 5 Keeping the torch's heat focused on the back side of the joint, touch the brazing rod to the other side. The rod will melt and capillary action will pull liquid braze to the source of the heat, filling the joint. Keep adding braze until the joint begins to overflow. It doesn't take much.

STEP 6 As soon as the braze is done, let it cool a little, then rinse with water and scrub lightly with a wire brush. Flux will eat the base metal if it's left on.

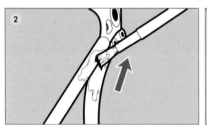 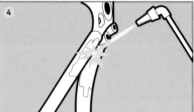 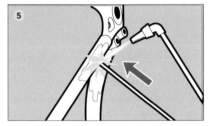

51 BEND METAL WITH FIRE

Metals are ductile—apply force and they bend. That is, of course, if you're dealing with thin metal—anything above ¾ inch (2 cm) in diameter requires hydraulic presses and stresses way beyond DIY. Or you can apply heat: Get even the burliest structural steel to red hot and it becomes a flexible plastic, bendable with easy pressure, maintaining its original strength when cooled.

STEP 1 Before applying heat, figure out exactly what you want the metal to do—hang a reference drawing where you can see it while working. Mark the bends on the stock with chalk or soapstone. **Pro tip:** Metal cannot compress down to zero—if you want a tight angle, the metal on the inside of the turn needs to go somewhere. Either make a slit before the bend with a chop saw or angle grinder, or accept that the curve will be gradual.

STEP 2 Fix the nonbending part in a vise, and clamp on pipes or square tubing to use as levers just past where the bend will go. The longer the lever, the tighter your control but the farther you'll need to move the torch.

STEP 3 Fire up the torch and liberally apply even heat—cold spots can tear or crush adjoining metal. If you have a rosebud tip, this is what it's made for. If not, use a cutting torch as if you're preheating for a cut, but turn down the oxygen so you can't accidentally burn the metal.

STEP 4 Heat until the bend area is a consistent cherry red, then put the torch down in a safe place but keep it on. Apply slow, steady pressure on the levers—it will get harder as the metal compresses and cools, so stop and reheat whenever the red color fades. Let it cool and repeat as needed for each bend.

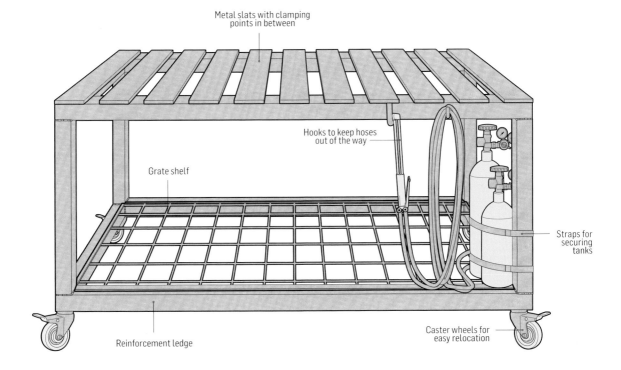

Metal slats with clamping points in between

Hooks to keep hoses out of the way

Grate shelf

Straps for securing tanks

Reinforcement ledge

Caster wheels for easy relocation

52 BUILD YOUR OWN WELDING TABLE

Two birds, one stone: This project is a solid beginning welders' craft and a must-have piece of gear for your work space.

STEP 1 Choose stock from metal suppliers—they'll be half the price of big-box stores—or scrounge up a metal bed frame in the trash. The whole table must be made from the same material, so choose wisely: 1-inch (2.5-cm) square tubing (thick wall is perfect, thin wall OK) or 1-inch (2.5-cm) angle iron are both widely available and will work. A sturdy table measuring 3 feet (90 cm) high, 6 feet (1.8 m) wide, and 40 inches (1 m) deep, will take 100 feet (30 m) of stock steel. Get naked steel—no paint or galvanization.

STEP 2 Measure from the floor to a comfortable working height and make it wide enough so you can reach across. Instead of a solid top, run stock from side to side in parallel lines, 4 inches (10 cm) apart, to give you clamping points. A shelf or reinforcement ledge near the floor are also helpful.

STEP 3 Determine the required number of pieces and their dimensions, then cut them all out in one sitting with a chop saw or grinder. Try to get as many pieces as possible out of your stock and remember to account for kerf.

STEP 4 Lay out the tabletop's outside edge on the floor, clamping all the components to something that's a known square—cement blocks and bricks are perfect.

STEP 5 Measure where the parallel crossbars will go, and draw marks on the outside edge for each crossbar. Turn on your acetylene and then slowly bump up the oxygen to 7 to 10 pounds per square inch (0.5–0.7 kg-force/sq cm). Use the torch to tack in a crossbar piece at the center, then two others halfway between the center and each end.

STEP 6 Stop. Walk away. Then come back later and look at your work with fresh eyes. Look OK? Then weld the tacked parts into place. Weld the edge pieces all the way around, paying attention to the inside corners. The crosspieces just need a weld on the top and bottom of the frame.

STEP 7 Tack in the rest of the crosspieces at their marks. Check for straightness, then weld them in place. Weld at one end, then the other, hopping back and forth to keep heat from building up too much in any one place.

STEP 8 Build the legs up from the top, then the crossbars for the supports or shelf. Keep marking, tacking, and walking away, then coming back to look it over again. If you made a mistake, cut the tack with your grinder and tack again. All welded as you like it? Flip it over and put it to good use.

53 CUT HEAVY METAL WITH FIRE

There's no better way to cut plate steel than with an oxyacetylene cutting torch—except technically it burns, the jet of oxygen igniting a line through the metal. The torch can cut through thicknesses and in positions that stymie even the toughest plasma cutter. Take note: The cutting torch only works on carbon steel. As a general rule, if it rusts, flame-cut it; otherwise, explore other options.

STEP 1 Use a cutting torch. Cutting torches are different from regular welding torches—they're longer and have a long lever on their belly. A regular torch just won't cut it.

STEP 2 Crank up the oxygen to 30 to 40 pounds per square inch (2–2.75 kg-force/sq cm)—way more than the usual 7 to 10 pounds per square inch (0.5–0.7 kg-force/sq cm).

STEP 3 Support your metal piece toward the middle and along its edges by propping it up on blocks. If you just support the edges, the piece will bow to the middle as the cut progresses, crushing melted edges.

STEP 4 Make sure you're cutting on clean metal—grind or sand away rust or paint, then scribe or soapstone where you need to cut. **Pro tip:** Keep your lines straight by setting up torch guides. Scrap lumber works but will get charred; bricks or metal guides work well. Remember to account for the width of the torch.

STEP 5 Practice your movements. Take the torch and pretend to make the cut. The torch head should be perpendicular to the metal, and move slowly and steadily along the cut. If it gets too awkward to move the torch, reposition yourself or break the cut into parts.

STEP 6 Prepare the gas flow. Make sure the valves on the torch are closed. Open the acetylene valve on the regulator (at 10 pounds per square inch [0.7 kg-force/sq cm]). Open the oxygen up to about 30 to 40 pounds per square inch (2–2.75 kg-force/sq cm).

STEP 7 Open the acetylene valve on the torch and ignite it using your sparker. You should get a feathery flame around 10 inches (25 cm) long, pluming black smoke.

STEP 8 There are two oxygen valves on the cutting torch: one at the base (near the acetylene valve), the other closer to the head. Open the lower one all the way, then slowly open the upper valve. The texture of the flame will change, getting brighter, shorter, and more focused. Keep adding oxygen until the flame is a circle of tight jets.

STEP 9 To test the cutting jet, squeeze the long handle on the torch body, releasing oxygen into the central hole on the torch head. You should see a needle of blue flame at the center of the circle and hear an intense hiss.

STEP 10 Hold the torch over the beginning of the cut, close enough so the circle of flame is just touching the metal. Preheat the metal until it glows cherry red. Make sure the face of the torch is perpendicular to the metal. **Pro tip:** Darkness helps. Cut in low light if possible, in the shade if not. You'll be able to see the color changes more easily.

STEP 11 As soon as the base metal turns red, squeeze the long handle. A stream of oxygen will jet into the red-hot steel, burning it in a straight line. You should see a stream of fine sparks shooting out of the bottom of the cut.

STEP 12 Slowly and steadily move the torch to cut the steel. If the cut stops and the metal just gouges, you might be going too quickly. If the steel turns from red to bright yellow and liquefies, you are going too slowly. Release the oxygen lever, skip ahead in the cut 2 inches (5 cm), and start again with preheating. Go back and cut the skipped portion after it has solidified.

STEP 13 When you reach the end of the cut, release the lever. If you need to go back to cut a skipped spot do so now, while it's still warm.

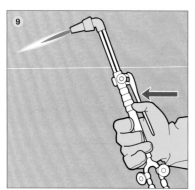
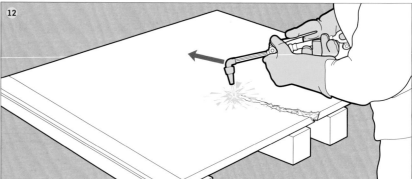

54 FLAME-CUT IN STRAIGHT LINES

Keeping the torch at the correct height can be difficult in long or complicated cuts, so use cheater wheels. Take a piece of scrap pipe or dowel that's 2 to 3 inches (5–7.5 cm) in length and attach a skateboard bearing to either end. (Use a pipe or dowel with a slightly larger outside diameter than the inside diameter of the bearing, and press them together in a vise). This is your axle. Mount the torch on the axle a couple of inches from the head, positioned so that the torch face is the proper distance from the metal you will be cutting—about ¼ to ½ inch (0.65–1.25 cm). When cutting, balance the torch on the wheels and use your hands for guidance.

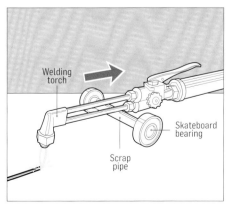

HACKETT SAYS

55 GO ON THE HUNT FOR OBTAINIUM

Like it or not, we live in a wonderful time of excess and waste: The stuff that past generations of engineers gave all to make is languishing in the trash. I call this stuff obtainium, but it is more an attitude than a class of substance—a way of looking at the world as a ripe thing, ready to be transformed into your vision. Learn to see as a scavenger and you'll make off with a ton of mighty steel, perfectly good wood, discarded electronics that still carry a charge, and plastic casings that can house whatever wild contraption you see fit to build. For free.

To start, travel to fringe areas where industry has languished and machines have gone to die. Transit centers are dumping grounds for unwanted-by-someone-else equipment and lightly used packing supplies, and universities and hospitals constantly leave barely outdated gear on the curb.

But watch out: Junkyards and failed industrial parks are crawling with scrapseekers who may not appreciate the competition. I advise you to steer clear of copper for this very reason—no pipe is worth a chance encounter with a person in need of a fix. Wear heavy boots and gloves, and get a tetanus shot (or three). Don't tread in places that are laden with glass shards, look structurally unsound, or otherwise set off sirens in your head.

Bring lightweight tools so you can harvest obtainium without alerting security. A multitool, a Phillips head screwdriver, a wrench, and a crowbar all fit nicely into an unassuming backpack. (Remember that even having a bolt cutter is illegal.) Be prepared to leave obtainium that you can't budge—or come back with a truck.

Why all this trouble? Think of something basic, something you have always depended on other people to provide: flour. You want DIY flour, and lots of it. For that, you need a grist mill—big wheels turning, powered by a pond. Huge stones are involved—you may just need to build a quarry. But then look at it through the lens of obtainium. What is the essence of a grist mill? Lots of force and the smooth and steady crushing of wheat. The stones are there for mass. What else has mass and can spin that smoothly? A few sketches and wandering around a car-part graveyard, and suddenly a salvaged truck axle is fixed to repurposed 55-gallon (208-L) drums of cement. And you have flour.

56 ASSEMBLE A PAINT-CAN FORGE

The stereotype of a blacksmith as a filthy, muscle-bound, crafty lunk is partially true. They were strong and dirty; the dimness is just a misreading of the partial deafness that years of hammering can cause. Blacksmiths were the inventors and hackers; for most of human history, they were technology.

A paint-can forge allows you to dip a toe into the world of blacksmithing without investing very much time, money, or space. It can be built with hand tools by almost any level of maker.

BUILD

STEP 1 Make the hole for the conduit connector—the part that will allow the blowtorch to pipe in flames—with a hole saw. Skip the sharp burrs and dented can by sticking scrap 2x4 in a bench vise, laying a small chunk of 2x4 on top, sliding the can over, and drilling down into the wood.

STEP 2 Screw in the conduit connector, nut on the inside.

STEP 3 Saw a firebrick in half horizontally with a ripsaw. The 9-by-4½-by-2½-inch (23-by-11.5-by-6.5-cm) brick should become two 9-by-4½-by-1¼-inch (23-by-11.5-by-3.25-cm) bricks (slightly less, accounting for kerf). If you have the half-size repair bricks instead of whole ones, skip this step.

STEP 4 Slather a layer of refractory cement inside the can.

STEP 5 As the cement sets, you're going to cut the bricks further down into trapezoids. Rectangular brick, round can—math to the rescue! You will need to compute two circumferences ($C=\pi d$): the circumference inside the can, and the inside circumference accounting for brick thickness. With a 6¼-inch- (16-cm-) diameter paint can, the inside circumference is 19 inches (50 cm). The bricks are 4½ inches (11.5 cm) wide, so you will need fewer than five brick halves. The bricks are 1¼ inches (3 cm) thick, so the inside circle will be around 4 inches (10 cm) in diameter with a 12½-inch (32-cm) circumference. Each brick's width should taper from 4 inches (10 cm) down to 2½ inches (6.5 cm). Mark the bricks accordingly.

STEP 6 Set the bricks on a stable surface. Rasp each brick into a trapezoid according to your measured markings. **Note:** The dust from firebrick will irritate everything it touches. If you wear contact lenses, flush your lenses and eyes out as soon as you are done.

STEP 7 The can top and bottom (now more aptly called the back of the forge and the door) need firebrick coverage, too. Measure your can's height and subtract your brick's thickness, then trim the trapezoids down to the remainder. Take care in cutting; you will need those drops of brick.

STEP 8 Drill a burner hole through one trapezoid. Break the delicate sides of the hole. We want this junction to be rugged and heat-trapping, with as much mass as possible.

STEP 9 Do a loose, cementless fitting of the bricks inside the paint can. It should be pleasingly snug. If there are any gaps larger than ⅛ inch (3 mm), carefully saw an assortment of shim stock from your drops of firebrick.

STEP 10 Spread an even layer of refractory cement around the inside of the can, no more than 3/16 inch (5 mm) thick. Lay brick chunks at the bottom, covering the circle as tightly as possible. Then, starting with the brick halves that straddle the burner hole, lay the trapezoids around the can's circumference. Fill gaps with trimmed brick scrap.

STEP 11 To place on the inside of the door, make a raised center of firebrick that will firmly slide into the forge. Don't make it too exact; brick-to-brick contact will crush its way to smooth.

STEP 12 Cut two circles of insulating fabric so they are slightly larger than the inside diameter of the ends. While you're at it, cut a strip slightly longer than the circumference. It should be wide enough to line the can. Cut a hole in this strip for the burner. Spread a thin, consistent layer of refractory cement. Lay in the fabric, squeezing and pressing out any bubbles with your fingers.

STEP 13 Bend a framework to support the forge. (Remember the old sawhorse? That design will do just fine.) Use the sheet metal screw in the conduit connector to lightly grip the propane torch. Keep pliers handy, or use scraps of frame material to make a door handle.

STEP 14 Let the cement cure overnight. Run the burner for a couple of minutes, let it cool, then run it for another 15 to drive out remaining moisture. Now all you need to do is blacksmith something.

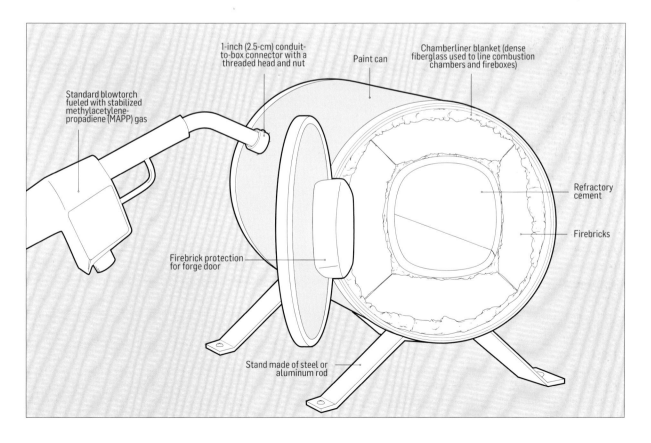

57 FORGE A KNIFE

Access a mental image of blacksmithing: Metal gets heated, hammered, ad then into a bucket of seething liquid it goes. When steel is cherry red, around 1,400°F (760°C), it reaches "normalization": It is just plain steel with wide-open potential. Hammering decides the metal's physical shape—in this case, a knife. Its final crystalline structure—whether it's brittle or strong, springy or stiff—is decided by the quench (the toasty oil bath you sink the metal in). Quenching in oil will make for a hard but nice and flexible blade.

STEP 1 Obtain a hunk of steel. A great knife steel can be found in leaf springs or files. Using a grinder or sander, remove anything that is not the knife, leaving the final edge at least $1/32$ inch (1 mm) thick. Drill any desired holes.

STEP 2 Choose a place with decent ventilation and a lack of flammable objects. Put the forge on a solid, level surface free of obstruction. Have fire suppression nearby.

STEP 3 Prepare the quench. Used motor oil is good, as is high-temperature vegetable oil. Pour the oil in a metal or ceramic quenching container deep enough to take the whole blade, point down, with about 2 inches (5 cm) for bubbling at the top. If the oil bursts into flame, cover it with a tight-fitting lid.

STEP 4 Fire up the forge. Leave the door off so you can monitor it as the heat climbs. Using pliers, lay the knife blank (the piece of metal that will become the knife) on the forge's bottom. Then wait the steel to get up to 1,400°F (760°C). This will take a few minutes after the interior of the forge is there.

STEP 5 Test for normalization. Since normalized steel is immune to a magnet's charms, grab the cherry-red blank with pliers and touch it to a magnet secured to a chunk of firebrick or another nonflammable object. If it sticks, put the blank back in and let it get hotter. If it does not stick, you are normal. Place the blank back in the forge to get back up to heat.

STEP 6 On an anvil, a piece of railroad track, or some other chunk of heavy steel, hammer your hot, red hunk of steel into shape. You will likely need to reheat the steel a number of times, as it will cool some while not in the forge.

STEP 7 Quickly and evenly, lower the knife point-first into the oil. If you see flames, release the blade. Grab the lid with pliers, cover the quench, and let it snuff.

STEP 8 When the blade stops glowing, pull it from the oil and clean it until it's shiny. Pop the knife into an oven preheated to 450°F (232°C) and let it cook for an hour or two. Leave it in long enough for the shiny metal to turn a straw yellow. Let it cool at room temperature. Now you are a blacksmith.

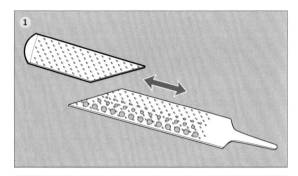

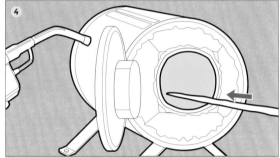

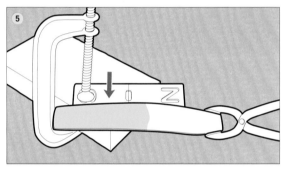

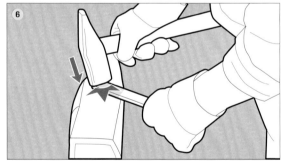

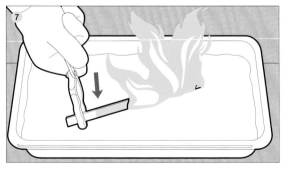

58 CAST A SHOT GLASS

The mini forge gets hot—hot enough to melt glass. Glassblowing is a deep and wide (and very dangerous) craft—far beyond the purview of this book. Casting glass is simpler. No handling of molten stuff is needed to produce solid, custom glass parts, like this shot glass.

STEP 1 Carve the mold. Mark a firebrick with two concentric circles, 1½ inches (4 cm) and 2 inches (5 cm) in diameter. Using a pick, knife, or file, carve out the space between the circles until you have a circle trench about 2¾ inches (7 cm) deep. Then carefully shave down the top of the center plug ¾ inch (2 cm). This will be the bottom of the glass, the trench the sides. If you want a witty saying about the dangers of drinking from amateur-made glassware, carve it on the plug now.

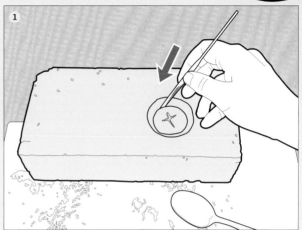

STEP 2 Smooth any sharp transitions with a file then dust the trench and the top of the plug with talcum powder; this layer will keep glass from sticking to the mold.

STEP 3 Fill the mold with crushed glass to the top of the brick and beyond. Clean, smashed bottles with labels removed will work. Since the glass will settle as voids collapse, make sure there is enough to fill.

STEP 4 Put the mold in the forge. Fire it up with the blowtorch.

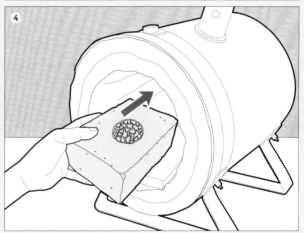

STEP 5 Using a high-temperature thermometer, monitor the temperature. When it reaches 1,550°F (843°C), let the glass bake for a half hour.

STEP 6 Take a quick peek at the mold. The glass should have melted, slumped, and smoothed. If it is not smooth, bring the temperature up another 50°F (10°C). Wait.

STEP 7 Once the glass is smooth, shut off the heat. Let it cool slowly, overnight if needed. **Note:** Glass is not very flexible. Heating or cooling it quickly can cause violent cracks and the possibility of shrapnel from thermal shock. Let it take its time.

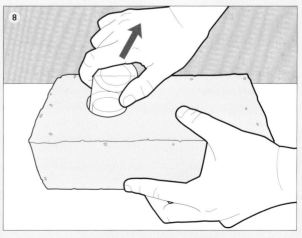

STEP 8 Flip the mold. Tap gently opposite the glass. You will have a weird, milky, possibly sharp-edged shot glass. Prepare a celebratory drink, read the ominous warning you carved, and decide to give away the glass as a gift, instead.

59 TRANSFORM A FORGE INTO A FOUNDRY

The first foundry I built was a couple of steel buckets, barbecue charcoal, and a vacuum cleaner. Total materials cost was US$6. I was melting aluminum an hour after I started the build. Also, after three pours, the foundry just fell to pieces. The next one was pretty ambitious. With a 55-gallon (208-L) drum, DIY refractory cement, and a powerful rocket burner, I was able to melt 40 pounds (18 kg) of bronze at a time and discovered what a catastrophic crucible failure looks like. Thankfully, if you built a forge already, all of the hard work has been done. To transform your forge into a foundry:

STEP 1 You need a template for the foundry's interior walls. Mark a 2-inch- (5-cm-) diameter circle in a piece of cardboard, then measure the inside diameter of the forge and make that circle concentric to the 2-inch (5-cm) hole. Cut the cardboard.

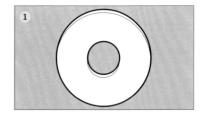

STEP 2 Mark the center of a paint-can lid, then cut out a 2-inch (5-cm) hole with a hole saw.

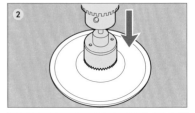

STEP 3 Make a handle for the lid using heavy wire. A steel welding rod measuring ⅛-inch (3 mm) will work, as will a doubled-up coat hanger. It should pass around the lid, avoiding the center hole, with multiple contact points on the lid itself. Make it strong. Attach the handle with ½-inch (1.25-cm) self-drilling screws, but only from the outside in.

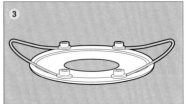

STEP 4 Make three cardboard tubes: one 2 inches (5 cm) in diameter, one with the outside diameter of the cardboard template, one with the outside diameter of the lid. The first two tubes should be slightly longer than 4 inches (10 cm), the third 2 inches (5 cm). Using the cardboard spacer, glue the tubes to the paint-can lid: the one that is 2 inches (5 cm) in diameter goes just inside the lid's hole, the middle one fits just inside the forge's interior, and the third goes onto the lid rim. All tubes should be on the lid's inside.

STEP 5 Prep the insulation. Carve a firebrick plug a little less than 2 inches (5 cm) in diameter and about 4 inches (10 cm) long. Set it aside. Saw and rasp firebrick to fit inside the cardboard layers as best you can. Fill gaps with refractory cement and rock wool.

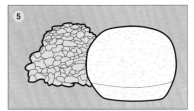

STEP 6 Fill the forms. Lay down a thick layer of refractory cement inside the lid. Jam firebrick chunks in, filling gaps with refractory cement. Ice with a final layer of refractory cement. Top with rock wool fondant. Let it set overnight.

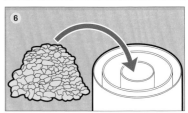

STEP 7 Fit the lid into the forge can. Cut off the cardboard with a utility knife.

> "So, you have a metal thing, but you would like another metal thing instead? With your own foundry, you can melt the first into material for the second."

60 CRAFT A CRUCIBLE (OR TWO)

The crucible holds the stuff that needs melting. Crucibles come in a wide range of sizes, are made of various high-temperature ceramics, and are absurdly expensive. A steel crucible will work well enough for this foundry, but it will eventually break down. I recommend you just build more.

STEP 1 Obtain two 2-inch (5-cm) black steel pipe caps. Using an angle grinder, grind down the ends of the caps until they are flat enough to sit inside your foundry without rocking back and forth. **Note:** Take care to get black steel—not galvanized, not stainless. Galvanized will outgas heavy metals on the first few firings; the poisoning will ruin your week. Stainless steel is oddly vulnerable to molten aluminum—the bottom will drop out the second time you use it, without warning (trust me). If only galvanized is available, soak the fittings in a 50/50 mix of muriatic acid (available in most hardware stores as "concrete cleaner" branded hydrochloric acid) and water for a few minutes, or until the bubbling stops. Use the leftover acid to clean your filthy sidewalk. Remember to neutralize the acid with a baking soda and water mixture before use.

STEP 2 Obtain a 1-foot (30-cm) black steel pipe nipple that is 2 inches (5 cm) in diameter and threaded on both ends. Put this pipe in a bench vise and, using an angle grinder with a cutoff wheel (or a hacksaw), cut the pipe in half, leaving two 6-inch (15-cm) pipes with one smooth end and one threaded end. Screw each of the pipe caps onto the threaded pipes. Measure the depth of your foundry and then subtract the thickness of the lid. Cut each of the crucibles-in-progress slightly shorter than that length, allowing them to sit inside the foundry without touching the lid.

STEP 3 Using a center punch, mark holes opposite each other, about ½ to 1 inch (1.25–2.5 cm) down from the uncapped end of each pipe. Using a ¼-inch (6.5-mm) bit, drill holes.

STEP 4 Obtain four ¼-inch- (6.5-mm-) diameter stainless-steel eyebolts, 2 inches (5 cm) long. Screw the eyebolts into each of the holes. **Note:** If zinc-plated eyebolts are the only option, strip the zinc with acid as before.

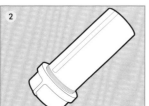

61 FASHION A CRUCIBLE CARRIER

Melting and pouring metal requires all of your concentration. A crucible carrier allows you to safely lift the crucible out of the foundry, while keeping your fingerprints intact.

STEP 1 Measure the midpoint of a 7¼-foot- (2.2-m-) diameter steel rod, as well as 1 foot (30 cm) out from the middle toward each end. Mark with a scribe or tape. Measure and mark at 2 inches (5 cm), 3½ inches (9 cm), and 5½ inches (14 cm) in from either end.

STEP 2 Secure the rod in a vise. Slide a 2-foot (60-cm) length of ½-inch (1.25-cm) pipe over the rod. Then, use your mechanical advantage to bend the rod perfectly. Crush the last few degrees in the vise.

STEP 3 Use the ½-inch (1.25-cm) pipe to make a series of matching bends. Bend a right angle at the 1-foot (30-cm) mark. This will be the handle. At the 5½-inch (14-cm) mark, bend the two pieces away from each another so they are at 90 degrees to the doubled rods and 180 degrees to one another. Then bend a 3½-inch (9-cm) right angle in each piece, but downward. The rods should be parallel but a little bit apart. Finally, bend both pieces up one more right angle at 2 inches (5 cm), straight up from the parallel bits, curving in a tight little arc.

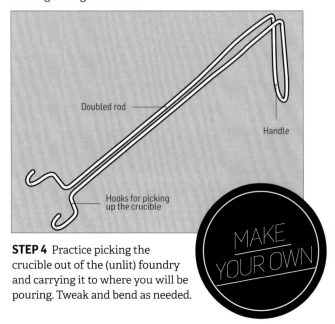

STEP 4 Practice picking the crucible out of the (unlit) foundry and carrying it to where you will be pouring. Tweak and bend as needed.

MAKE YOUR OWN

62 ASSEMBLE A CRUCIBLE POURER

When pouring molten metal, you will need as much control as possible while maintaining a livable distance from the little cup of closed casket you have brought into the world. To assemble:

STEP 1 Drill two through holes in the (nonthreaded end) of a 3-foot (90-cm) length of ¾-inch (2-cm) steel pipe.

STEP 2 Bolt a 2½-inch (6.5-cm) stainless-steel muffler clamp onto the newly drilled holes. Connect the 3-foot (90-cm) length to the downstroke of a ¾-inch (2-cm) T connector. Screw two 6-inch (15-cm) nipples of ¾-inch (2-cm) steel pipe in the other holes.

STEP 3 Leave the pourer right next to where you will be pouring. Set the hot crucible into the muffler loop. Lift the crucible with the pourer, keeping your dominant hand about one-third of the way down the pipe, holding it level with one of the 6-inch (15-cm) handles. To pour, use your dominant hand as a pivot point and turn the handle to tip the crucible. Control is the most important thing; practice all you need to.

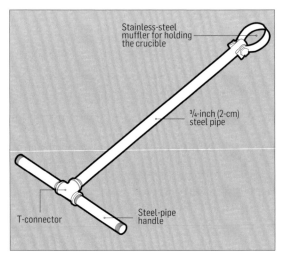

63 HAMMER A DROSS SKIMMER

"Dross" is the scum that rises to the top of molten metal—heat-strained garbage that has been purified out through the magic of fire. Before pouring, you need to skim off the scum. To build:

STEP 1 Obtain a 3-foot (90-cm) length of ¾-inch (2-cm) pipe. Tighten the pipe into a vise.

STEP 2 With a hammer, smash the last bit of pipe as flat as you can, then use it to skim off the nasty dross that gathers in your forge. **Pro tip:** The word "dross" is awesome. Using it in casual conversation will make you sound like a turn-of-the-20th-century preacher with a dark, twisted, secret life.

64 REFINE ALUMINUM

I do not drink beer and think soda tastes like robot sweat, but cans can be recycled, maker style. Be careful: When molten metal touches anything that has the tiniest residual bit of water, the water turns to steam, cuing an explosion.

STEP 1 Crush a lot of cans—more than you would think—then load the crucible with as many as possible, using a hammer and a rod to crush them. Put the crucible in the foundry, plug the hole, and fire it up.

STEP 2 After 10 minutes, peek into the viewing hole (wear sunglasses or welding goggles). If all you can see is a smooth layer with gray bits, congrats: You have melted metal. Feed in cans until the crucible is full, then replug.

STEP 3 After the new cans melt, clean off the top layer with your dross skimmer. Repeat skimming and feeding until there is only 1 to 2 inches (1.25–2.5 cm) of free space inside the crucible. Then open the foundry lid and, using the crucible carrier, set the crucible down in the pourer loop.

STEP 4 Sprinkle a fine dusting of unscented talcum powder (which is resistant to high heat) on a steel muffin tin. Pour molten metal into the tin.

STEP 5 Let the tin cool. These metal muffins can be the charge for your next pour—the ingots that you'll melt for future adventures in casting.

65 FOCUS ON MASONRY

Metal might hold our world together, but masonry *is* our world. From flint axes to ceramic insulators, stone and its variants have been used by makers since we got the gift of thumbs—it was our first raw material. Romans dreamed up concrete, Egyptians baked mud into bricks, and the Maya carted limestone. You can follow in their footsteps—or roll a wheelbarrow down the same belabored path—and harness the hard stuff for your own projects.

CEMENT VERSUS CONCRETE

Pretty much everyone gets this wrong: Cement is the powdery mix of limestone, calcium, silicon, iron, and aluminum that, once combined with water, gets poured into molds. Meanwhile, concrete is made up of an aggregate (like sand, gravel, or crushed stones), water, and cement itself. When these ingredients are placed in a mold, the latter two mix, forming a glue that holds all the aggregate together.

WHEN WORKING WITH MASONRY, WEAR A RESPIRATOR TO KEEP DUST OUT OF YOUR LUNGS.

IT'S DURABLE. ONCE THAT MIX SETS, IT IS NOT GOING ANYWHERE. ON THE DOWNSIDE, UGLY LANDSCAPE FEATURES ARE NOW FOREVER.

IT'S FORMABLE. POUR, CAST, MOLD—WE CALL THE MASONRY INTO BEING. IT DOES WHAT WE SAY. UNTIL IT SETS…

IT'S CHEAP. DIRT IS USUALLY FREE, AND CEMENT DOESN'T COST A WHOLE LOT MORE. EVEN WITH A TINY BUDGET, YOU CAN STILL MAKE SOMETHING THAT WILL DRAW GAWKERS IN A THOUSAND YEARS.

IT'S A GREAT INSULATOR. All other areas of making benefit from masonry's energy-sucking density and internal structure. Sound waves, light, vibration, electric current, heat, or hard radiation—if you want to stop stray energy, masonry is the way to go.

CEMENT DOESN'T DRY— **IT CURES.**

SAND IS PRETTY MUCH VERY TINY STONES. IT IS A CRUCIAL INGREDIENT FOR BUILDING MATERIAL AND MAKES A GREAT ABRASIVE WHEN GLUED TO PAPER OR BLASTED VIA AIR. IT CAN ALSO BE MIXED WITH OIL TO MAKE CASTING MOLDS OR MELTED INTO GLASS.

STONE IS QUARRIED AND CUT, OR GATHERED ROUNDED FROM A RIVER. IT LASTS FOR THOUSANDS OF YEARS. **PRO TIP: NEED ROCKS? ALL YOU WILL EVER WANT, FREE!** JUST GO STRAIGHT DOWN FROM WHERE YOU ARE RIGHT NOW. SOME DIGGING MIGHT BE REQUIRED. WATCH OUT FOR BURIED POWER LINES; NYC RESIDENTS, BEWARE OF CHUDS.

BRICKS ARE BAKED CLAY— HARD AND BRITTLE. They're useful for building anything that requires solidity. Can be stacked naked for short lengths and nonstructural use, or glued together with mortar to make things as big as buildings.

Plaster-of-paris (gypsum plaster) is the most common type of store-bought masonry. Just add water and you initiate a chemical reaction that reforms gypsum from powder into a paste—soak it in gauze and you have a cheap, easy casting material that can replicate faces and splint limbs.

THE HOLES IN CINDER BLOCKS AREN'T FOR CLASSY GOOD LOOKS: These hollow spaces (called cores) keep the blocks lightweight for transportation and basic building. Once they're in place, plug them with cement and rebar to reinforce your structure.

GLASS COMES FROM SAND.

66 MIX UP A BATCH OF CONCRETE

Concrete is artificial and man-made—something that is easy to forget in the cities and sprawls where anything green seems like sloppy editing—but composed of natural components. Details differ for different uses, but the basic recipe has worked since Roman times.

STEP 1 Use a hoe to mix dry ingredients—cement, sand, and gravel—in a 1:2:3 ratio. Begin with sand and cement until they are thoroughly blended, then add gravel in batches. Make sure the mix is consistent. **Pro tip:** Instead of obtaining a large container, just mix on top of a tarp.

STEP 2 Add water in small doses, mixing as you add. The mix should be just pourable.

STEP 3 Decide what needs concreting, but decide quickly—after two or three minutes, any concrete left sitting will start to cure into a solid that will outlive you. Every 1 square foot (929 sq cm) of 4-inch- (10-cm-) deep concrete requires 50 pounds (23 kg) of concrete mix.

STEP 4 Use forms. Shockingly, concrete—a metaphor for heaviness—is heavy. Until it cures, it will want to slump. A form keeps your sidewalk, driveway, or durable altar tight.

STEP 5 Scrape the top of the pour with a board, pushing down. Gravel will sink, and the skim—a smooth cement layer—will rise. Smooth with a trowel, then inscribe for posterity.

67 MAKE A SOLID BASE FOR A STRUCTURE

The solidity of a structure is directly proportional to how well it is tied in to the gold standard for solidity: the Earth. If you want your gazebo to remain while the rest of the trailer park blows away, or your guerrilla art to still be there when you get out on parole, anchor the posts in concrete.

STEP 1 Lay out your posts. Some kind of plan is helpful but not mandatory.

STEP 2 Dig some holes. Lots of holes. More holes than you think necessary. **Pro tip:** Up-to-code standard is a hole three times the diameter of the post, depth one-third the height of the post.

STEP 3 Raise, erect, and level the posts. Keep them upright and centered with the stakes and wire anchored outside the hole.

STEP 4 Ensure the ground connection. It is important that the post contact the ground directly for moisture transfer, and to prevent rot and rust. After the post is raised and centered, shovel a little dirt in around the outside of the post and tamp it down.

STEP 5 Pour. Fill those holes with concrete.

STEP 6 Cure. Keep the supports up for a few days—thick concrete can take some time to cure.

68 BUTTER A BRICK

Applying mortar looks easy as smearing butter on toast. And with a little bricklaying practice, it can be.

STEP 1 Trowel grip is crucial. Hold the handle firmly with your thumb extended along the top instead of wrapping it around the side.

STEP 2 Grab your hawk (a handy platform with a handle—a scrap of wood will also work fine) and load it up with mortar, then load your trowel by slicing off a V-shaped wedge. **Pro tip:** Position your trowel on its edge and use it to cut through the mortar like butter. Scoop mortar onto your trowel with a quick twist of your wrist, then snap or flick the trowel to create suction and keep the mortar from sliding off.

STEP 3 To lay a mortar bed for bricks, drag the trowel along its side on the ground, then use the tip to draw a central line down the mortar. Make furrows along this line to grip the bricks.

STEP 4 Place your first brick on the mortar and check that it's level. Then butter your next brick: Mix your mortar with quick chops, then pick some up on the trowel. Smear it across the brick end, then apply two quick dashes of mortar at either end.

STEP 5 Place this brick next to the first one and use your trowel to pat it level, then scrape the excess mortar off the side. Keep buttering those bricks and laying them down.

69 CAST A PLASTER MASK

Plaster is another form of dehydrated rock, using gypsum where cement uses limestone. Infusing fabric with plaster provides a structural integrity and easy, detailed molding.

Quick to set and light in weight, it's often used in making casts—but you can also use plaster-infused bandages to make masks.

STEP 1 Prepare the victim. Take your extra head and coat it with Vaseline, using a heavy hand on the hair parts.

STEP 2 Mix the plaster (on the watery side) in a wide container. Cut 18-inch (45.75-cm) strips of gauze, and mix them in the plaster.

STEP 3 Prepare the life-support system. Cut a couple lengths of straw, insert them in the victim's nostrils, and then tape the straws in place.

STEP 4 Squeeze out excess plaster from the gauze strips, and cover the head with bandages (gauze + plaster = bandage). Do two layers, offsetting the second 90 degrees from the first.

STEP 5 It should take 20 minutes to firm up. **Pro tip:** Leave without telling the victim. **Pro pro tip:** Make noises as if you were leaving—push your chair back, open and close the door—but stay in the room, silently.

STEP 6 Use scissors to carefully cut the cast, going from ear to ear over the top. Carefully pull the cast away, leaving the head behind.

STEP 7 Put the cast halves back together, and repair the seam with a plaster strip. It should be solid in an hour.

STEP 8 For a creepy, lifelike mask, paint the inside face part with two-part latex. Let it cure, give it some paint on the outside, cut eyeholes, and dip a toe into the exciting world of armed robbery without a care.

70 SET UP A GLASS WORK SPACE

Glass is the ultimate embodiment of itself. Hard, brittle, and smooth, glass does not possess these characteristics; it is these characteristics. Humans and glass go way back, from unbelievably sharp obsidian (volcanic glass) blades to the classiest drinking vessels. It confounded and killed our ancestral enemy, the dinosaurs (currently in bird form); makes an excellent insulation; and is easily available everywhere. If you are going to be working with glass, prepare beforehand to save time later.

SOFTEN YOUR SURFACE Tape a couple of layers of cardboard onto your work surface. It could save your work if knocked over.

LIGHT IT UP Fire a light across your work surface at a low angle. Any stray bits of glass will sparkle, giving away their location.

SWEEP FOR METAL Work spaces are constantly littered with screws, bits of wire, and chips of metal. Since metal scratches glass easily, do a sweep. You'll save yourself polishing time.

PAD YOUR CLAMPS Obtain a 6-inch (15-cm) segment of bike inner tube. Stretch a couple of inches over the jaw of your clamps. Fold the remainder down to make a pad, securing it to itself with electrical tape. Pull off the tubing and use it whenever glass needs clamping.

CHIP AWAY Work with glass and you will make chips—tiny razors that climb into your flesh and never leave. Use a maker lint brush—duct tape, sticky side out, wrapped around a scrap of wood—to clean up. Roll the wad along flaking scores and irregular seams every time you see a glint.

71 MOUNT GLASS

In general, mounting glass involves filling a loose-fitting connection with a flexible sealant. Glass mounting systems are all designed for inevitable failure. It will all fall apart, they say, but it need not happen catastrophically.

MIND THE POINTS Small, pointy bits hold glass in position, centered in an oversize frame (the smallest pin nails work well). If the points fail and the glass and frame can move independently, the possibility of both surviving remains. Still, so does the probability of broken glass.

FILL THE GAP Something soft (compared to the frame and glass) and flowable—caulk, putty, molten lead—runs into the dead space, sealing out the elements. **Pro tip:** You might not want the squish of caulk, especially near skin or things that will be assembled or reassembled repeatedly. Bike inner-tube rubber makes a cheap, durable rubber gasket. Cut a strip of rubber long enough to fit around the glass and wide enough to come right under the edge of the frame (double if necessary to fill the gap). Overlap edges 1 or 2 inches (2.5–5 cm). Use a tube repair kit to secure gasket-to-gasket ends. Once it cures, you have a perfect, custom-built gasket.

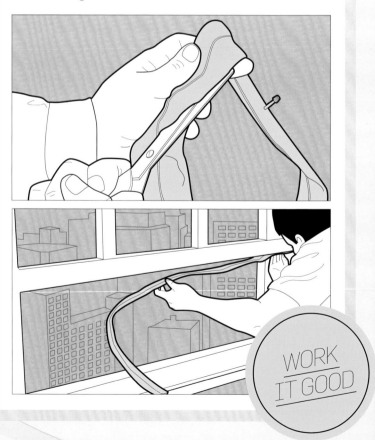

WORK IT GOOD

72 CUT GLASS BOTTLES WITH STRING

You may have heard about "cutting" a glass bottle by soaking a string in fuel, tying the string around the bottle, and setting the fuel on fire. That method has never worked for me, but it's possible to "cut" a bottle by carefully scoring and heating it in just the right way. "Cutting" is the wrong word for this process; what's really going on is a kind of controlled shattering. It may take a little while to get the hang of it, but bottles are free for the taking everywhere, and you can always recycle your mistakes.

STEP 1 First you'll need a nice, even, continuous score line around the bottle's circumference where you want the break. This line is made with a glasscutter's wheel, which should be available at any hardware store. You'll also need a jig to get a straight line that doesn't wander. Jigs for bottle cutting are easy to build and inexpensive to buy; poking around online should turn up lots of ideas and options. Set the glass to the cutting wheel and score the bottle all the way around in one continuous operation. Resist the temptation to go back over your score line to make it deeper or touch up any mistakes.

STEP 2 Now heat up a pot full of water to near-boiling, stand over a sink with a folded towel in the bottom, and pour hot water over the score line as you rotate the bottle to evenly distribute the heat. After about 10 seconds, switch to the cold-water tap, rotating the bottle as before. You may hear small popping sounds as the glass begins to crack along the score line—hold the bottle up to the light, and you should be able to see the break. Repeat the heating and cooling process until the top of the bottle falls free into the sink.

STEP 3 Flatten the sharp cut edge by lapping it in circular motions against a piece of scrap window or mirror glass dusted with silicon carbide sandblasting grit and a little water. Finally, round off the corners with a piece of wet-or-dry sandpaper until they feel smooth to your fingertip.

73 PICK PLIERS

Pliers are used for grabbing and manipulating things that are too small, hot, sharp, or otherwise beyond the pay grade of fingers. They can be used as a brute-force tool for wrenching, prying, and starting the seized, or as a precision tool for picking up parts fat fingers might drop.

LINEMAN'S PLIERS Originally used by electrical linemen for gripping, bending, and cutting cable. Sturdy, with nonconductive handles and cutting jaws that can chomp through screws, nails, and most wire. Some sets are even built to withstand a certain voltage.

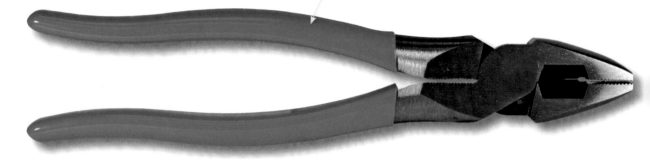

WIRE STRIPPERS Exceedingly handy for electrical work. From the cheap, adjustable ones to fancy, gauged, and automatic ones, they work a lot better than your teeth when it comes to stripping, cutting, and bending wire.

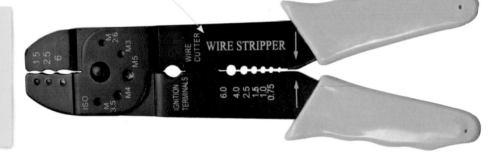

SLIPJOINT PLIERS Its adjustable fulcrum allows you to grab larger things while maintaining the same angle and distance between handles—increasing the size of your grip, but without increasing the size of your hands.

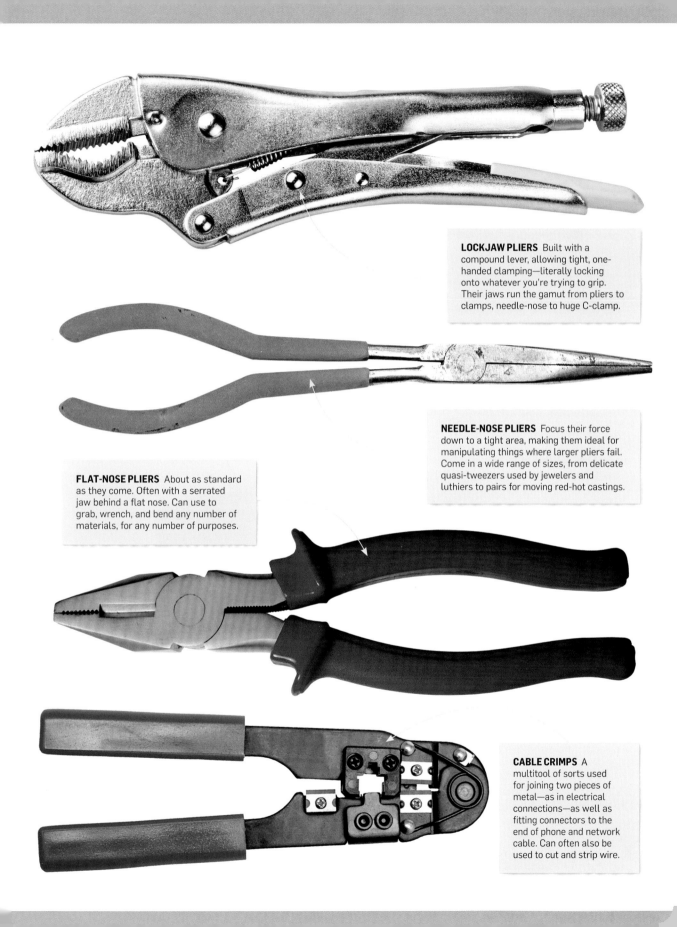

74 TACKLE TAPES

If "stickiness" is the manifestation of the van der Waals forces, with stacked opposing molecular magnetic fields grabbing one another, think of tape as sticky rope—the strength of the bond is limited to the strength of the tape. Use it to connect material faces together along the same plane, to seal low-pressure holes, and as a lash only in times of desperation.

DUCT TAPE Originally intended for repairing HVAC ducts, duct tape has become a symbol of ingenuity and sketchy work. While its strength comes from the thread that runs the length of the tape, its quality is all over the place. Use it any time you want people to think you don't care. **Pro tip:** Become a duct tape snob. Good duct tape is rated for things like temperature range, tensile strength, and corrosive resistance.

GAFFER TAPE Gaffer tape is like duct tape but made with nonreflective fabric coating. Used widely in the film industry, it's still held to professional standards. Although it can be shockingly expensive, use it if you want something to work.

MASKING TAPE Masking tape is meant to mask, covering areas to remain unpainted or sandblasted. While it's not intended for joining, it will hold paper together.

CLEAR PACKING TAPE Used for sealing cardboard boxes as well as a quick-and-dirty laminate, clear packing tape makes an airtight seal. **Pro tip:** On every roll, tape some lightweight insulated wire to the center tube. Keep the wire parked under the cut end of the tape, eliminating the frustrating game of "who has fingernails?"

VINYL TAPE Also known as electrical tape, vinyl tape is electrically insulating, strong, and robust. A surprising amount of the underground electrical infrastructure is insulated with tape.

75 CHOOSE AN ADHESIVE

Glues, epoxies, spit: Adhesives work beneath our notice, at the microscopic level, latching on to surface imperfections we can't see and creating plastics before our ignorant eyes. There are two broad types: evaporative, where evaporating solvent leaves connective material, and reactive, where a chemical reaction leaves a plastic behind. **Note:** Most decent glues will give you cancer, but only in California. Many residents weekend in Vegas to get crafting projects done safely.

RUBBER CEMENT Rubber cement is essentially latex dissolved in solvent. When the solvent evaporates, a layer of rubber remains. It's good for waterproof connections and ones that need to flex. **Pro tip:** Electronics projects can be weatherproofed with rubber cement, but make sure heatsinks and vents remain uncovered.

WATER-BASED GLUES From Elmer's white to wood glue, water-based glues work when water evaporates, leaving a bond. They are the linchpin of industries from woodwork to macaroni-bedazzled Mother's Day cards. **Pro tip:** The tighter the clamp, the better the bond.

CYANOACRYLATE Better known as superglue, this stuff gets into pores and crevices to bond (and also makes excellent sutures). Expose the glue to moisture—even just low-humidity air—and a chemical reaction occurs, forming a plastic. While it makes a great bond, it's very brittle, and its reaction is exothermic to the point of possibly igniting materials like cotton and wool.

TWO-PART EPOXIES The strongest adhesives, two-part epoxies work by mixing two chemicals to create a reaction that leaves a plastic or resin behind. Most bonds are somewhat brittle, but some construction epoxies can be stronger than cement. Like superglue, reactions are usually exothermic and can be intense.

76 LOCK (AND UNLOCK) THAT THREAD

Threaded connections are satisfying to create, and rated fasteners tell you exactly how strong the link will be. Unfortunately, nothing is perfect, and chaos sneaks in through tiny flaws and vibrations, slowly urging rated nuts and bolts to part ways. A little glue can add the extra level of connection needed to keep it together. The most common brand of thread-locking adhesive is Loctite, but other adhesives will work as well in most cases.

STEP 1 Make sure the connecting hardware is clean, with no cutting oil and no stray metal chips.

STEP 2 Start the threaded fastener as you normally would. Once the first few threads have engaged, run a short line of thread locker up the fastener. The longer the line, the more secure the thread will be.

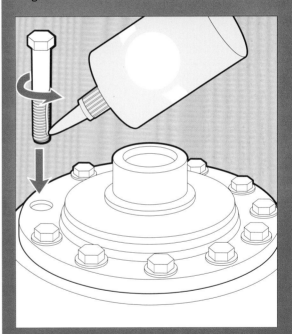

STEP 3 Connect the hardware, and wipe away excess glue before it sets. **Pro tip:** Thread locker sets up quickly—even quicker if you're using superglue.

STEP 4 When you need to break the brittle connection, a sharp shock at the beginning of unscrewing will help; persuade, then unscrew. Clean the last bits of thread locker with a wire brush before reconnection. A quick soak in acetone can speed the process.

77 FOCUS ON PAPER AND CARDBOARD

With a little science and brute force, things in nature can become useful. Crush some trees, cook them up, add bleaching agents, and strain, and suddenly filthy bird ghettos are transformed into a blizzard of fresh white sheets—potential by the truckload. Cardboard is paper with thicker stock and a sine wave of corrugation; layer it in a 90-degree laminate and its strength approaches that of plywood. Both materials are infinitely useful.

FOR MOST OF US, THE LARGEST WORK SPACE WE HAVE IS THE FLOOR.

Working sprawled across the floor is liberating for the mind, crippling for the knees. Wad cardboard, secure with tape, and make yourself custom kneelers. To create a huge, resilient work surface for large projects (and to protect the floor), lay down two or three layers of cardboard.

FISHMOUTH MADE EASY.

ROUND PIPES ARE EXCEEDINGLY HARD TO FIT UP, ESPECIALLY WHEN MORE THAN TWO COME TOGETHER IN A FISHMOUTH. FIND TWO CARDBOARD TUBES A LITTLE LARGER THAN YOUR PIPE STOCK, THEN HAVE AT THEM WITH A BOX CUTTER UNTIL THEY FIT WITHOUT GAPS. SLIDE THE CARDBOARD TUBES OVER THE PIPES AND USE THE CARDBOARD AS AN EDGE TO GUIDE YOUR SCRIBE.

MAKE BOLD LISTS ON PIECES OF CARDBOARD.

MAKE TEMPLATES PRINT YOUR DRAWINGS ONTO PAPER, GLUE TO MULTIPLE LAYERS OF CARDBOARD, THEN CUT OUT THE THICK, DURABLE TEMPLATE. IT WILL BE AS TOUGH AS A PLYWOOD TEMPLATE BUT WITH WAY MORE FLEXIBILITY FOR TRANSFERRING TO IRREGULAR SHAPES. **PRO TIP:** FOR REPETITIVE PARTS OR SHAPES, LEARN THE OFFSET OF YOUR CUTTING TOOLS AND ADJUST YOUR DRAWINGS BY THIS AMOUNT (USE THE "OFFSET PATH" FUNCTION IN DRAWING SOFTWARE). THEN PRINT OUT THE TEMPLATE—YOU NOW HAVE A TOTALLY CUSTOM, REUSABLE CUTTING GUIDE.

Layer newspaper on your table, cover with lots of cardboard, and you can now cut or paint or solvent with impunity.

CLEANLY CUT PIPES.

Cutting pipe that's larger than your pipe cutter becomes a game of chance. Paper can save the day. Obtain a sheet with at least one factory edge, at least 2 inches (5 cm) longer than the pipe's circumference. Wrap it around your pipe so its edge is continuous, and scribe around it, then cut with a saw or torch.

MAKE PAPER MODELS.

Good drawings are key to building anything more complicated than a sandwich.

I keep a rotation of notebooks, hopping from one to the other as I move between projects. At a glance I can see my thought processes unfold, problems get solved, or crucial mistakes occurring right there on the page. Many makers have ditched old-school notebooks in favor of high-grade 3-D software (see entries #173–175), but I find that my ink sketches and scrawls always translate better into big or complicated. And if you make with what you have, not what the plan calls for, software will often frustrate.

78 MAKE A SCREEN FOR SCREEN PRINTING

There are precious few places where all the makers of the world come together, and screen printing is one of them. Circuit boards, broadsides, T-shirts, etched dies—if you want to slap a detailed, consistent, repeatable image onto something, screen printing is the way to go. At the core of screen printing is the screen itself—a very fine woven mesh that supports a stencil (a negative of your image). Start now, and you could be completing professional prints by this time tomorrow.

STEP 1 Screw or glue together wood as a frame. Make it close to the size of your intended print. **Pro tip:** Extra punk rock points for using an old picture frame.

STEP 2 Staple the screen mesh to the top of the frame. Pull the mesh tightly against the frame with no creases or folds, stapling every 1 inch (2.5 cm) or so until it's taut. Trim the excess mesh. **Note:** Mesh is rated by number—the lower the number, the coarser the image—and can be found at craft stores or online. Mesh rated 125 is fine for posters and line art, while 225–300 is best for small text and circuit boards. **Pro tip:** More punk rock points for using gauzy curtains that you found somewhere for free.

STEP 3 Tape where the mesh and wood overlap. **Pro tip:** Pre-stretched frames are about US$20.

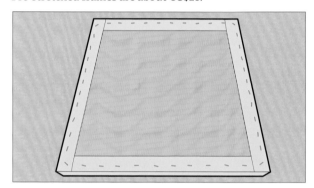

79 BURN AN IMAGE ONTO A SCREEN

Once you've tacked and stretched (or bought) your screen, all that stands between you and printing out that stack of flyers is an actual image to print. Put your image of choice onto the screen like so.

STEP 1 Measure your image against the screen. If there's extra space on the screen, tape it up. If you are using a transparency, your image needs to be dark, dark, dark. **Pro tip:** If your printer can't get dark enough, print two or three identical transparencies, then layer and tape them.

STEP 2 Prepare a photo emulsion according to the manufacturer's instructions. Then run a line of emulsion across the tip of the screen, and use a squeegee to evenly coat it. Get the screen into darkness and let it sit as long as the instructions say.

STEP 3 Place the transparency or cutout on the emulsion. Lay a sheet of glass or acrylic on top.

STEP 4 Let the light hit it. The emulsion will change color when it's exposed (check specific product for instructions). Bright, artificial light will harden the

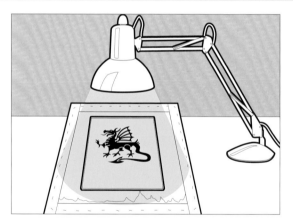

emulsion slowly (20 minutes to an hour), while UV (black light) does it quickly (40 seconds). **Pro tip:** Sunlight does a good job (20 seconds) but is hard to aim and goes from exposed to overexposed quickly.

STEP 5 Stop! Push the lamp away, or persuade the Moon God to eat the Sun. Once your emulsion becomes overexposed, there is no un-exposing.

STEP 6 Rinse. Unset emulsion will wash away, and your image should remain. If everything washed away, it was underexposed. If not enough came off in the wash, it was overexposed. Do it again, but better.

80 SCREEN-PRINT BY HAND

Many screen-printing businesses will make a custom screen print for under US$50, a lot less if you're willing to take a used screen. On your own, here are the steps.

STEP 1 Prepare the press. Put your backing board on the bottom, the printee in the middle, and the screen on top. Adjust until everything is lined up as if you were printing. Trace the outside of the screen on the canvas.

STEP 2 Build the press. Using shims and tape, make a positioning jig for the printee on the board. Run little tabs of tape to grab any corners that are not being printed on. Position a hinge at the top and attach.

STEP 3 Load the press. Put the printee in place with the screen tilted up.

STEP 4 Close the press. Run a line of ink across the top edge of the screen, then grab the squeegee with both hands and firmly pull toward your body while applying pressure down into the screen. Do another pull, just to be sure that the design is well coated in ink.

STEP 5 Lift the screen and pull out your print, setting it aside to dry. Repeat to print as many copies as you like, then wash your screen—ink dries fast and can ruin it. Take that, Warhol.

WORK IT GOOD

81 ASSEMBLE A MAKER'S SEWING KIT

Attaching things is fundamental to making, and a sewing kit is an important part of any maker's arsenal. You never know when you'll need to patch a torn tarp—or perform hasty self-surgery.

NEEDLES Sewing needles are numerically graded—the higher the number, the finer the needle. Sharp needles are best for natural fibers like cotton, while blunt-edged needles are better for knit fabrics. Leather and sailmaker's needles have a triangular point designed to pass through suede, leather, canvas, or vinyl without tearing the material.

THREAD Cotton or polyester quilting thread is stronger than machine thread and perfect for hand sewing heavy-duty projects. Waxed nylon sewing thread works well for outdoor gear since it won't tear or rot in the rain. **Pro tip:** Dental floss works just the same.

CLOTH MEASURING TAPE A cloth measuring tape makes measuring oddly shaped items (or people) easy. **Pro tip:** In its absence, use a length of string and a standard measuring tape or yardstick.

STRAIGHT PINS Fabric has a way of shifting while you're sewing. Pinning it in place takes a few minutes, but will save you a lot of frustration. **Pro tip:** For thick leather or fabric, try small clamps or alligator clips.

SEAM RIPPER Even the pros make sewing mistakes, and a seam ripper is the easiest way to undo them. **Pro tip:** In a pinch, use a knife or a razor blade.

SCISSORS Any sewing kit is incomplete without a good set of sturdy scissors. Save your teeth the trouble.

82 KNOW YOUR STITCHES

Different needs call for different stitches. Start with threading a needle and knotting the end of your thread, and end by sliding your needle under the last stitch and tying the thread around itself with an overhand or square knot. In between, let your project choose the stitch.

PICK YOUR STITCH		
RUNNING STITCH		The most basic of stitches, used for joining two pieces of fabric together. Pull your needle through the fabric, moving in and out, ultimately ending with a stitch resembling a dashed line.
BACKSTITCH		This quick stitch creates a seam that rivals the security of machine sewing. Make the first stitch as you would a running stitch. Then double back through the hole that completed your first stitch, creating a solid line.
OVERCAST STITCH		A sturdy stitch used to finish the raw edges of a seam. Starting 1/4 inch (6.5 mm) from the edge of the fabric, stitch from back to front by looping the thread over the fabric edge, making your stitches evenly spaced.
CATCH STITCH		This stitch works well for making a strong, flexible seam. Start by making a backstitch on one fabric. Then cross over the stitch and seam, making another backstitch on the adjacent fabric.

83 CREATE QUICK AND DIRTY PATTERNS

Sewing a square patch on a flat surface is one thing, but what about making a 3-D object like a gear bag or welding sleeves? Most professional patternmaking involves math and computers. One quick solution is to sacrifice an old T-shirt or a holey gear bag, cut it at the seams, and use the pieces as a pattern.

STEP 1 Cut along the seams of your sacrificial bag or garment, breaking it into its constituent parts. Each pattern piece should lay flat on your new fabric so that it's easy to trace and cut. **Pro tip:** It helps to iron pattern pieces before tracing them, and numbering them with a marker can help you avoid unnecessary seam ripping.

STEP 2 Using these pattern pieces, cut the shapes out of whatever material you're using for your new bag or shirt or welding sleeves. Connect them at key points using straight pins or Chicago screws.

STEP 3 Connect the dots along the seams using a needle and appropriately weighted thread. **Pro tip:** If sewing takes too long, you can cheat by using a bunch of evenly spaced Chicago screws.

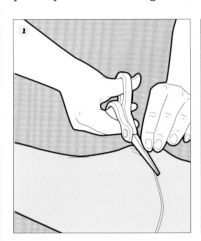
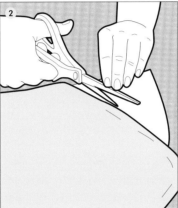
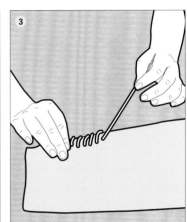

84 WIN AT LIFE WITH NO-SEW FASTENERS

When thread won't hold, no-sew fasteners can help you keep it together.

SNAPS Snaps are the Velcro of metal fasteners; they create relatively strong but impermanent connections between materials. A snap has four parts—both the ball and socket each have a small metal ring with spikes that pierce through the fabric to secure them. Snap setters and snap pliers help keep snap parts in place while you hammer them down; they're available at most craft stores.

CHICAGO SCREWS Chicago screws (a.k.a. screw posts and sex bolts) are some of the finest fasteners ever created, and essentially work like removable, adjustable rivets. Each slotted screw has a corresponding threaded post, creating a tight closure that can be loosened with a flat-head screwdriver. To install, simply punch a hole through both pieces of fabric, insert the screw into one piece and the post into the other, and screw them into place.

GROMMETS Grommets come in handy when repairing tarps or shower curtains. Use them any time you need to create or reinforce a hole so that it won't stretch when pulled. You'll need a hammer, a base tool, and a flaring tool (make sure you get the same size tools as your grommets).

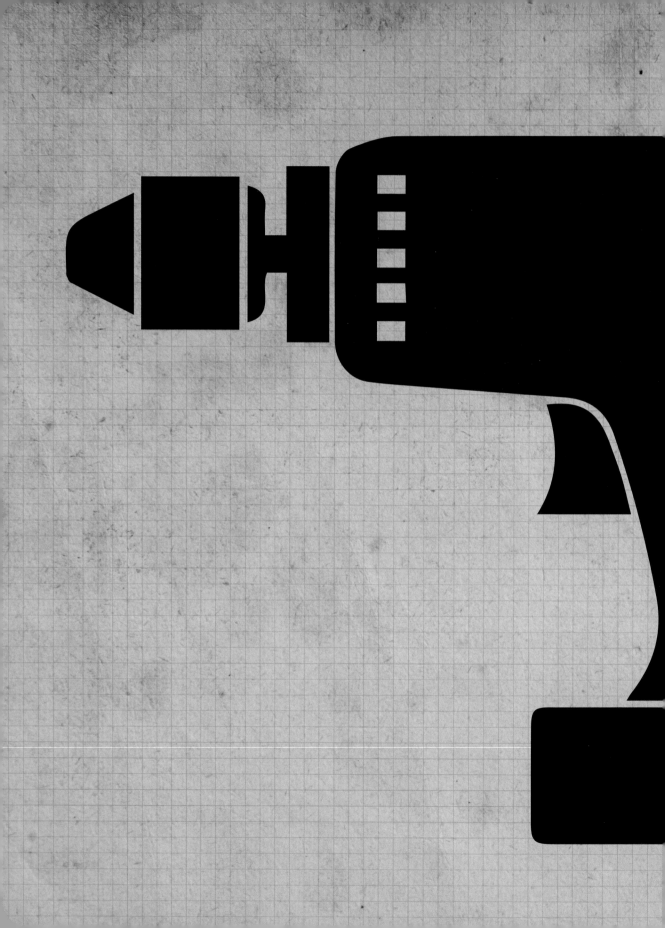

85 PICK A POWER SAW

Where the power of handsaws is limited by the power of your hands, these saws get their juice from longer-lasting sources like batteries or a wall socket. Letting them do the hard work can save you a lot of pushing and pulling.

CIRCULAR SAW Essentially a motor, saw blade, handle, and base plate with a guide. Versatile cutting tool, as long as you're cutting in a straight line. Choosing the right blade for the job is essential.

RECIPROCATING SAW Most often used in construction, remodeling, and demolition. Cuts with a blade that moves back and forth. Good for cutting numerous materials in hard-to-reach places.

BANDSAW Consists of a strip of toothed metal passing over two wheels, which feed the material under the blade. What it lacks in portability, it makes up for in utility, cutting curves and straight lines through materials thick and thin.

PORTABLE BANDSAW Works well for smaller projects, cutting through anything from pipes to rebar, from plastic to wood. Some come in variable speeds for greater control and cut accuracy.

JIGSAW Something of a tiny reciprocating saw. Can be used to cut straight lines, but really intended to cut curves. The thinner and narrower the blade, the tighter the curve.

BENCH SAW Also called a table saw. It's basically a circular saw that protrudes up through a table or bench. It comes with adjustable blade height and angle, making it great for long cuts.

86 MAKE MASTERFUL CURVES WITH A JIGSAW

Excellent for custom parts and stencils, horrible for sticking to a straight line, the power jigsaw dates from the 1940s. Jigsaws ride light on top of the work, anchored only by the oscillating blur of the blade and you.

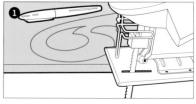

STEP 1 Plan your cut. Curves, stencils, oversize novelty glasses—transfer your design to the work surface with a writing utensil. **Pro tip:** Three teeth should be contacting the work at all times. Make thin materials like sheet metal thicker by firmly attaching them to plywood backing—spray adhesive will work well.

STEP 2 Don't start the saw under load. Hit the material with the blade moving. To start an internal cut, drill holes slightly larger than the blade. Make the holes inside where the cut will go, and all evidence of them will disappear.

STEP 3 Continue working, keeping close to your drawn edge. Until your skills increase, don't try for perfection on your first cut. Remember, you can always cut more—less, not so much.

STEP 4 Break complicated cuts into smaller ones. Some tight turns are impossible—saw blades only cut in one direction.

STEP 5 Keep it tight. Jigsaws work best when all of their energy goes into the teeth hitting the material. Any looseness wastes energy and breaks blades. Push firmly, clamping as much as possible.

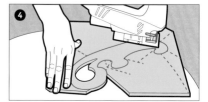

87 CHOOSE A CIRCULAR SAW BLADE

Simple and casually dangerous, the circular saw is the angle grinder of wood. Use the correct blade to leverage its might.

A CROSSCUT Very fine and very slow, crosscut blades carve across the grain. Creates a smooth cut, sans unsightly splinters or tears.

B HOLLOW-GROUND Almost as fine as crosscut, the body of a hollow-ground blade is thinner than its teeth, making it faster.

C PLYWOOD Plywood blades are slow and have many fine teeth. For finished plywood, cutting sometimes takes a while. **Pro tip:** You can slice aluminum and other soft metals with any of the finer blades (crosscut, hollow-ground, plywood), but keep it cool and steady. Otherwise, the aluminum version of sawdust will be melted by the friction and eventually clog up the saw.

D RIPSAW The name says it all. With fewer teeth, a ripsaw blade is great for rough cuts with the grain. It won't create a quality finish, but it gets the job done fast.

E COMBINATION Designed to both crosscut and rip, combination blades work slightly more slowly and leave wood slightly prettier than ripsaw blades.

F ABRASIVE An abrasive blade will grind its way through masonry and tile, and it's perfect for steel. Don't use it on softer metals—it will clog.

G DIAMOND Diamond blades claim to be for glass and ceramic, but making dry cuts in glass is asking for trouble. Use ample oil and beware flying shards.

88 SLICE WITH A CIRCULAR SAW

Most electric tools have a manual ancestor, and the circular saw's predecessor is the water-powered sawmill. While a lot of things are worth learning on your own, the ones that can cost you fingers or limbs are sometimes best read about.

STRAIGHT CUTS ONLY The saw blade is the only circular thing about this saw. Try to cut a curve and get a free lesson in the power of friction.

ADJUST THE SHOE The saw rides on the "shoe"—the plate at the saw's bottom. Adjust the shoe angle for beveled cuts. For cleaner cuts, set the blade depth as close as possible to the thickness of the material.

EVERY SAW IS A SNOWFLAKE Every saw (even within brands) has different lines and guides on the shoe to indicate where you're actually cutting. Do some test cuts when using an unfamiliar saw. **Pro tip:** Measure the distance between the blade edge and the side of the shoe. Write this on the saw with permanent marker or a label maker. Take some nail polish and draw a red arrow to mark the side of the blade. If it's a borrowed saw, the owner will thank you.

USE A GUIDE If you're using a circular saw, there's a good chance straight lumber and clamps are nearby. Position the lumber, clamp it down, and just ride the shoe against the guide.

WORK WITH A PORTABLE BANDSAW

The portable bandsaw is primarily used on job sites for tasks like trimming rebar or conduit during construction. Imagine a hacksaw and an arm that never tires—just a little too powerful, teetering on the edge of power tool and lawsuit. To use:

SECURE THE WORK Portable bandsaws pack some serious torque. If the thing you're cutting can move, lock it down in a vise or clamp it to something solid.

POSITION THE SAW AND YOU The bumper (flat sheet metal plate where the back side of the blade goes into the saw) is where all the real action happens. The work is jammed against the bumper, the only way out being through the blade. Make sure the bumper is in a position to do its job; otherwise, the saw will jump violently when blade touches work.

ELIMINATE THE LEVER The lever is a wonderful tool—except when it's you. The shorter the lever, the more control of the saw you have, so keep your arms as short as possible. I hug the saw to my chest like a vicious little baby.

BUCK IT Tubing, big stuff, timber—if you're cutting two sides at once, rock the saw back and forth, slowly, as you would with a chainsaw.

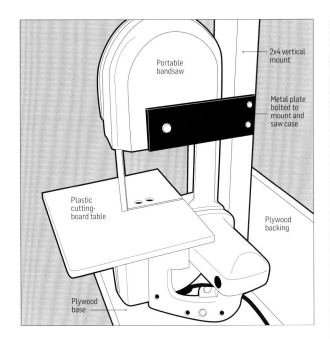

90 BENCH-MOUNT A PORTABLE BANDSAW

The portable bandsaw is powerful enough that domesticating it by locking it down seems benevolent—like capturing a feral dog once close to hunting humans, now happy on a leash. Since it's heavy and puts out a ton of torque, it needs to be locked down without the freedom to flex—but with the freedom to change blades as required.

STEP 1 You need an independent frame to support the saw in its new orientation. Here, two scrap pieces of ¾-inch (2-cm) plywood were wood-glued together into a butt joint, forming a stand. A 2½-foot (76-cm) 2x4 post provides a vertical support for the saw, but measure your own model to arrive at the correct length and width. Try adding wood spacers at the bottom to keep the saw from shifting.

STEP 2 Make a template for the support piece that will attach the bandsaw to the 2x4 post for extra security. With the saw mounted on the stand, look for hardware on the saw that could be removed and used to fasten it to a support piece, which in turn would attach to the 2x4. Trace the hardware locations, then use the template as a guide when cutting the support piece out of scrap metal.

STEP 3 A cardboard template will again come in handy when making the insert table for your newly secured bandsaw. Cut a square of the cardboard and make a slit in it to accommodate the saw's blade, then trace where the bolts will go to hold the table in place. Finally, transfer the template to a table material of your choosing. **Pro tip:** A plastic cutting board is a pretty perfect table material.

91 GET INVENTIVE WITH A BANDSAW

The portable bandsaw can do all the things the floor model can, and a bunch it cannot. While both can cut curves, the portable can cut them in more dimensions.

Ⓐ FLIP IT If cutting curves or shapes in flat material, flip the saw upside down, trigger on top. Push the blade through the work like the world's most powerful jigsaw.

Ⓑ MAKE RELIEF CUTS To get perfectly tight curves, you might need to make relief cuts. Cutting curves in shallow chunks can take some pressure off your blade. The smaller the chunks you remove, the tighter the curve.

Ⓒ CURVES IN ALL DIMENSIONS Most saws are rooted to standard x-, y-, and z-axes. The portable bandsaw makes its own planes, depending on how you hold it. Cut complicated, three-dimensional shapes by envisioning them as planar cuts through stuff rotated any way you want.

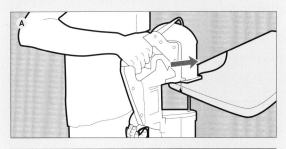

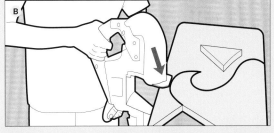

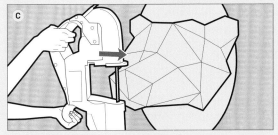

HACKETT SAYS

92 RESPECT THE ANGLE GRINDER

Recently, I was giving an artist a crash course in shop tools. Soon we got to the angle grinder: "Incredible power tool. Does it all. Full of love, but without compassion—this attachment here can sever your soul...." After what seemed like a moment (but I suspect was far longer), I noticed "the look": interest in the topic but with concern, confusion, and fear. I get this look more often than one would like. So, as a way of letting her know it was perfectly safe, I fired it up. "It's like a handheld shop!" I said, demonstrating some basic functions before turning it off and letting the motor wind down. Before handing it over, I mentioned how it could seem scary, maybe intimidating, but how it was really more useful than terrifying...and the grinder snapped on and spasmed in my hand, the cutoff wheel carving a deep gouge across three fingers. "And best of all, it cauterizes as it cuts!" I concluded, as she gawked at the still-smoking track across my hand. That led nicely to a tour of the first aid kit.

More than a handheld shop, the angle grinder is handheld making. It is simple but rich with potential: Learn the basics in 10 minutes, spend the rest of your life mastering it. Treat it well and you can do almost anything; take it for granted and it will take some fingers.

I have deep feelings for the angle grinder. A god in the old sense—powerful and helpful, but terrifyingly crazy in a way that reminds us of the new world's pecking order. It cares not for me, or any of us.

93 PICK A SANDER

Lose the tie before picking up a grinder or a sander. While sanders tend to be tailored toward wood, and grinders built to work on heavier materials like metals and concrete, both involve abrasive materials spinning fast enough to make you regret that rattail for a new reason.

ORBITAL SANDER A tool for finishing, on the spectrum between belt sanding and hand sanding. Great for prepping walls for paint or refinishing old furniture. Can finish coarse or fine, depending on the sandpaper you choose.

BENCH BELT SANDER If a bench grinder is a wheel, a bench belt sander is a tank tread—moving a single loop of sandpaper quickly in one direction. Mounted to a bench, this model can sand pretty much whatever you hold against it.

HANDHELD BELT SANDER Often used in early sanding stages, since it can remove a lot of material quickly. **Pro tip:** Hang on tightly—it tends to grab surfaces and can take you on a ride.

DIE GRINDER/ROTARY TOOL For small-scale cutting, grinding, sanding, polishing, and engraving. Has about as many functions as it does attachments. Can be surprisingly powerful, despite its small size and sound.

ANGLE GRINDER Has many uses, from finishing rougher metal pieces to tearing through the doors of a burning garage during a rescue operation. Ranges from glorified sanders to honorary chainsaws, depending on disc size, type, and the power of the motor.

BENCH GRINDER Essentially a whirring wheel used to grind, shape, or sharpen metal. With a brush attachment, you can use the same machine to polish or remove rust from tools, screws, and pretty much any metal implement.

94 DO IT ALL WITH AN ANGLE GRINDER

The angle grinder is the simplest of power tools: It's just a motor spinning with a disc, with no speed control or reverse. It cuts, grinds, shapes, and polishes, usually accompanied by a spray of photogenic, badass sparks. If you're interested in working with metal and have the budget for two tools, make sure one is an angle grinder. And if you have the budget for only one tool, well, make sure it's an angle grinder.

STEP 1 An angle grinder can and will hurt you in a very wide variety of ways. Burn, gouge, blind, deafen—grinders do it all. Always use safety glasses, gloves, ear protection, and a very healthy respect for the tool.

STEP 2 The grinder delivers a lot of rotational energy and will launch surprisingly heavy things. If the work weighs under 100 pounds (45 kg), clamp it down.

STEP 3 Use the side handle. Every shop has its own derogatory name for it, but the side handle allows the stability and control a beginner needs and an expert appreciates. Also, it makes it easy to shoot sparks at whoever mocks you.

STEP 4 Take a moment to see how the switch works with the grinder unplugged, as every grinder brand has a its own switch configuration and a different series of small movements required to turn the grinder on.

STEP 5 Plug the grinder in and fire it up, but never start under load. The centrifugal force of the spinning wheel will make the grinder jump; a firm, steady hand is needed.

STEP 6 Grind at the work. A shockingly large percentage of experienced metalworkers forget the "angle" part of "angle grinder." This refers to the angle of attack—the grinder body should be at a 45-degree angle to the work, and only the rounded edge of the wheel should make contact with the item you're grinding.

STEP 7 From beginning to end, start safe and finish safe. The grinder will shoot forth a plume of sparks and swarf. Be mindful of where they go—they can and will set things on fire, including you.

WORK IT GOOD

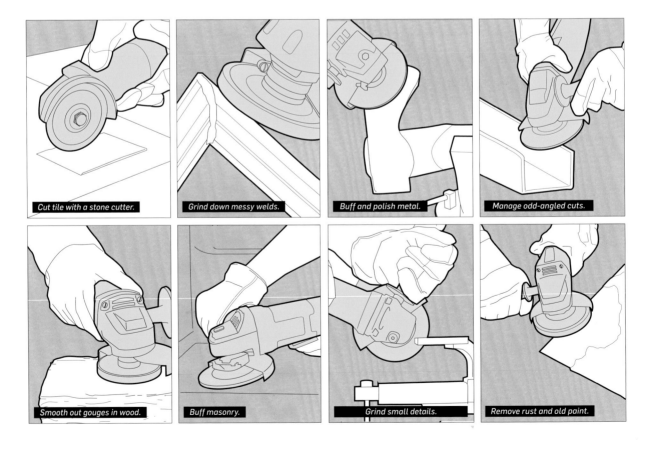

Cut tile with a stone cutter.

Grind down messy welds.

Buff and polish metal.

Manage odd-angled cuts.

Smooth out gouges in wood.

Buff masonry.

Grind small details.

Remove rust and old paint.

95 PICK A WHEEL, ANY WHEEL

The angle grinder is a pretty dumb tool—its real power is in the wheels. Swap out discs and the tool is totally transformed.

Ⓐ GRINDING DISC Use this disc to remove metal rapidly, which will leave a coarse finish. The standard size is about 4½ inches (11 cm) in diameter, but it goes up to 9 inches (23 cm) for the big boy grinders, at about ¼ to ⅜ inch (6.5–9.5 mm) thick. **Pro tip:** New grinder wheels have a crisp 90-degree edge that's great for notching tubing and cutting shoulders to look milled. I use new wheels solely for the crisp edge and grind with older wheels.

Ⓑ SANDING DISC You can mount any grit of sandpaper to this disc, turning your grinder into an essential tool for blending welds, subtle shaping, and all kinds of surface treatment. It can also be used on wood and plastic. It usually requires an adapter.

Ⓒ CONDITIONING DISC These come in a range of color-coded grades used in a wide variety of ways—from grinding down welds to polishing car wax. It connects to the base via Velcro. Watch out for the conditioning disc—as its base get banged around, the Velcro's hooks and loops tend to fail.

Ⓓ WIRE WHEEL There are two basic types. Knotted wire is thicker, more aggressive, and good for ripping out rust or paint; it'll leave a coarse pattern. Crimped wire, on the other hand, is much less aggressive, good for cleaning, and great for blending tool marks and scratches. **Pro tip:** Cheap wire wheels shed directly into you at about 117 miles per hour (188 km/h), usually at crotch level. Use this information as you will.

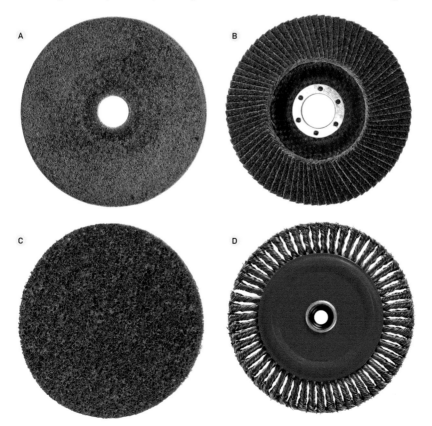

96 USE A CUTOFF WHEEL

The cutoff wheel is a tool to be respected and feared. Used properly, you can make complex 3D cuts in metal that are impossible to re-create with anything short of a CNC mill. Used sloppily, it'll maim you in ways that'll have ER doctors crowding in to take cell-phone photos.

STEP 1 Mount the cutoff wheel, which is much thinner than other wheels. Often, you'll need to flip one of the nuts to catch the inside of the wheel.

STEP 2 Break down complex cuts into a series of straight cuts and remember to account for kerf. If you need to cut curves, make them as shallow as possible—the wheel will explode at the slightest insult.

STEP 3 On longer or curved cuts, take the time to lightly score the whole shape, creating a superclear path to follow.

97 DRILL BETTER HOLES

Drilling a hole is a basic maker skill, so basic that there's little, if any, training. Still, the "hand" in "hand drill" renders it imperfect and prone to error.

STEP 1 Sharpen all of the bits you plan on using.

STEP 2 Secure the work. Eliminate all sources of shake. Machine tools are mostly solid cast iron for a reason. **Pro tip:** Always secure the work in at least two points.

STEP 3 Measure precisely. Chalk or marker aren't the best tools here. Sharp pencils are good; scribes are better.

STEP 4 Punch it. The tip of a drill bit will want to flex and wander, especially on smooth materials. Use a hammer or spring punch to keep the bit from skating off before cutting. **Pro tip:** Use a center drill or fat countersink to start holes. The dense tip will not flex or wander.

STEP 5 Use squares on the drill's side and front. **Pro tip:** For precision, avoid muscle memory. Make it abstract by flipping it upside down, or turn it into a series of tasks to concentrate on by using offhand drilling.

STEP 6 Move burrs past the item you're drilling by clamping sacrificial material to the bottom of the work. In most cases, scrap wood will do, and any burrs will gather there instead of in your hole.

98 DRILL GLASS WITHOUT A CRACK

Rare among materials, glass is completely human-made. Natural glass is exceedingly rare, and this output of volcanoes and lightning strikes bears scant resemblance to everyday glass. While people have been making glass for thousands of years, optically clear, cloudless glass became cheap and easy only recently. Now that we have it, let's drill into it.

STEP 1 Get the right bits. Diamond-coated or carbide-tipped glass bits should do the trick.

STEP 2 Glass needs to be secured way more than wood or metal. Make sure the piece cannot spin, and eliminate potential flex with wood, plastic, or layered cardboard shims. **Pro tip:** Avoid metal-on-glass contact. If using metal clamps, tape a thick layer (½ inch [1.25 cm]) of cardboard into the jaws.

STEP 3 Lay paper or gaffer tape where the hole will be, on both sides of the glass. You'll be drilling through the tape, and it keeps the glass surface from spalling (chipping or flaking).

STEP 4 Flood it. Temperature spikes from drill friction will crack glass as quickly as a suicidal robin. If possible, submerge the whole piece in water and drill it there, or get a helper with a spray bottle to keep you flowing.

STEP 5 Let the drill do the work. Don't apply force to it; the tape will keep the bit where you want it. Start slow and stay slow.

STEP 6 You should have a clean hole with no cracks. Use wet superfine sandpaper to remove shards from the hole.

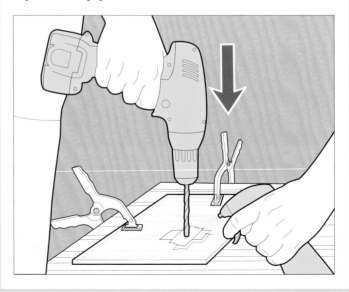

99 HANDLE A HOLE SAW

The term "hole saw" is simultaneously misleading and precise. It's precise because the name describes it perfectly—it saws a hole. It's misleading because it's used with a drill. With sizes ranging from ½-inch (1.25-cm) to "no-way-is-that-safe" gouges, hole saws must be respected. They are the best way to drill (cut? saw?) holes above ¾ inch (2 cm) in diameter.

STEP 1 Choose your saw. There are four types. Carbon steel is the cheapest and works best for wood and soft things (it'll shatter if used on metal). Bimetal works for metal and hard plastic (or slowly cutting wood). Tungsten carbide is great for fiberglass and quickly ripping through wood. And diamond-tipped is your saw of choice for glass and masonry.

STEP 2 Bring extra torque. Hole saws save effort by removing the bare minimum of material. However, that big circular blade reverses the usual mechanical advantage of a drill; the torque needed to punch through will stall small cordless drills and threaten wrist breaking with corded ones. Peck your way out. Keep the drill running, but dip it in and out of the cut; it'll cut slower but won't stall out.

100 CHAIN-DRILL SLOTS

That is not a drill you're holding; it's a mill seed. Unconstrained to the z-axis, milling machines are drill gods that can cut precisely in any direction. While drills only cut down, remember, "down" is decided by whoever clamps the work. Secure work sideways, upside down, or at an angle, planning carefully and clamping securely, and your drill gets closer to a mill. Sure, it'll be slow and the level of precision is in your hands, but if a mill can do it, a mill seed can, too.

STEP 1 Draw out your cut on the work, from a simple slot to a complex array of panels and switches.

STEP 2 Pixelate. Imagine the drawing was done in paint stick: What kind of pixel size would you need to accurately create the image? Straight runs could be a row of ½-inch (1.25-cm) overlapping dots, while tight corners might use smaller pixels.

STEP 3 Optimize. All of those sketched pixels are going to be "drawn" by you, using a drill bit as ink. Determine the smallest number of holes that need to be drilled.

STEP 4 Drill away. For clean slots, the holes should overlap as much as possible.

STEP 5 Use a file or a die grinder to clean up any jagged edges or irregularities.

101 PICK A DRILL BIT

Drill bits are tools used to cut, bore, and grind your way through a variety of hard materials—at least they are once they're attached to a drill. The tougher the material to be drilled, the tougher the bit for the job.

SPADE BIT Flat, with one centering point and cutters on either side. Most often used for rough boring in wood, causing the fibers to splinter upon the bit's exit.

REAMER Used for slightly enlarging and cleaning up already-drilled holes. More or less picks up where another drill bit left off. From spiral to straight, tapered to adjustable, there are a bunch of different types and sizes.

TWIST DRILL BIT The most commonly used bit. Essentially a metal cylinder with an angular point and grooves that spiral along its length. Different materials require their own angled points—generally, the softer the material, the sharper the angle.

HOLE SAW Essentially a cylindrical saw blade used for cutting holes. Apart from the material destroyed by the blade itself and the pilot bit sometimes attached, drills without eating up most of the core material.

COUNTERSINK BIT Used for drilling conical holes. In wood, handy for creating holes for screws or bolts to sit flush with or below the surface of an object. Can also help remove burrs from already-drilled holes in metal.

STEP DRILL BIT Conical and consisting of a series of steps. Can drill holes of various sizes, or enlarge or remove burrs from holes already drilled. Some smaller bits are self-starting, but larger versions may require predrilling.

GLASS DRILL BIT Where metal and wood are pretty good about not cracking or exploding when you drill into them, glass can be a little more finicky. These bits have a special carbide or diamond tip that, with patience and lubrication, can grind its way through.

MASONRY BIT Used, often with a hammer drill, to drill into brick, concrete, and some types of stone. An extra-hard tip allows the bit to burrow into exceptionally hard surfaces, while spiral grooves pull pulverized material out of the hole.

102 KNOW YOUR ROTARY TOOLS

The rotary tool is the simplest of all power tools: Attach a collet to a motor shaft, stick a tool in the collet, spin the motor, and with the whine of a thousand ravenous mosquitoes, you have a pocket machine shop.

This pocket machine shop will take geological amounts of time to do any tasks that are larger than pocket size. However, for small, detailed jobs—or if you're a maker with more time than money—the rotary tool can get you on the path to serious making with minimal investment.

TYPES OF ROTARY TOOLS

HANDHELD, AC-POWERED The temptation is to refer to all models of this tool as a Dremel, the most common name brand. But there are many, many brands of plug-in rotary tools, and all corded rotary tools are pretty much the same. The key difference is in how the tools are charged, so go with whatever you prefer or is cheapest.

BATTERY-POWERED Some take standard rechargeable batteries, while others have a proprietary battery pack. The higher the voltage, the higher the RPM with less gearing to break. Also, the battery will die more quickly than you thought possible. **Pro tip:** A rechargeable will take a different battery than all your other cordless tools.

FLEXIBLE SHAFT A big, burly motor runs a flexible shaft that spins the rotary tool bits. It's far, far stronger than the rotary tools you might be used to, with torque that can hurt you. This is a serious shop tool for fine, detailed work. It's hard to find new, but if you ever come across an older one, snatch it up and you'll never use the standard handheld again.

DIE GRINDER A rotary tool on steroids. It's not as maneuverable or user-friendly as a common rotary tool, but it gives you real power. It uses larger (usually no less than 1/8 inch [3 mm]) tooling and is better suited to serious metal- and woodwork than detail work.

AIR ROTARY TOOL Runs off compressed air and is usually found in light industrial and automotive settings. Air motors have the same high-speed, minimal torque of a corded rotary tool but with a lot more ominous whine. Great if you have a serious air compressor, incredibly frustrating if you don't. Usually lacks even the most rudimentary speed control and will stall from a strong look.

103 TRANSFORM A ROTARY TOOL INTO A TINY DRILL PRESS

Rotary tools are rich with potential, but they're only as precise as your shaky hands. Reduce the human factor and they can approach the precision of a drill press; get ambitious and you can make yourself a tiny milling machine. **Note:** Rotary tool drill-press attachments are available in stores, but you can make this one as precise or as quick-and-dirty as you need.

MATERIALS

- Junked dresser
- Saw
- Scrap 2x4 (around 2 feet [60 cm] will do)
- Drill
- Screwdriver
- Pipe hanger
- Hardware (wood screws, bolts, and nuts that fit pipe hanger holes)
- Bike inner tube
- Small bench vise

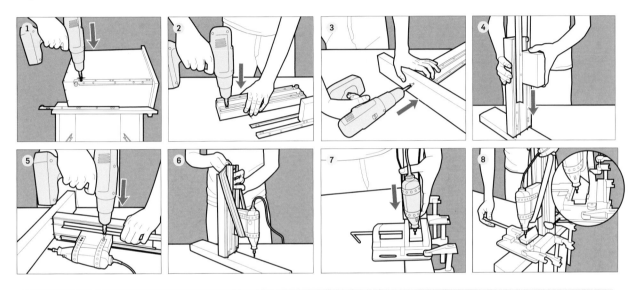

BUILD

STEP 1 Find a discarded dresser on the street somewhere. Pull a drawer from the dresser and remove the slides and guides—the slides will be on the drawer itself, while the guides will be on the inside wall of the dresser, or occasionally reversed. There should be two of each.

STEP 2 Attach the guides, parallel to one another and about 2 inches (5 cm) apart, to a piece of 2x4. (For extra obtainium points, use the side of the drawer instead.) Attach the slides to about 2 inches (5 cm) of 2x4, parallel and the same distance apart as the guides. Test the fit by running the 2x4 slider along the guides.

STEP 3 Attach the guides perpendicular to a base—make one with 2x4 or drawer parts, at least 18 inches (1.2 m) long. Check and double-check that the slides are 90 degrees to the base, then screw it all together.

STEP 4 Mount the 2x4 slider on the guides. It should move smoothly up and down.

STEP 5 Use the pipe hanger to attach the rotary tool to the slider, using at least two points of contact. Space the contacts out as much as possible for stability.

STEP 6 Attach a doubled-up length of inner tube between the rotary tool and the top or the guide support. It should be short enough to keep the tool tip a little off the base and elastic enough to make sliding the tool up and down easy.

STEP 7 To test the press, stick the vise right under the tool and clamp it to the base. Load it up with something that needs cutting. Fire up the rotary tool and ease it down the slides. Drill a hole, letting the inner tube reset the tool.

STEP 8 Bonus mini mill: Attach the work to the moving jaw of the vise. Position the rotary tool with a cutting bit (not a normal drill bit) right next to the work. Use the vise handle to move the rear jaw and the work through the cutter. It will only be one axis, so spin the vise 90 degrees for a new cut.

104 FOCUS ON METAL

Metal holds the world together. This is literally true—go straight down in the Earth, and you'll hit a giant dumpling of metal (iron and nickel mostly, with a healthy bedazzling of gold and platinum), surrounded by an eddying soup of the same, but liquid. This is also true in the day-to-day sense: Our civilization is built on metal foundations, crafted by metal tools, rewarded at the end of the day with aluminum cans of beer. Manipulate metal, and you manipulate the world.

THE OVERWHELMING MAJORITY OF ELEMENTS (91 OUT OF 118) ARE METALS.

EXOTICS, LIKE TITANIUM AND MAGNESIUM, ARE HARD TO SCROUNGE OUTSIDE OF AIRPLANE JUNKYARDS AND OFTEN REQUIRE LOTS OF VERY SPECIFIC KNOWLEDGE.

THERE ARE FOUR MAIN METAL TYPES.

CUPRIC COPPER–BASED METALS, SUCH AS BRONZE AND BRASS, ARE EASY TO SHAPE AND GREAT FOR ENVIRONMENTS THAT TYPICALLY RUST STEEL. THEY ARE ALSO GOOD CONDUCTORS THAT CLEAN UP NICELY. HIGH SCRAP VALUE; SCAVENGE AT YOUR OWN RISK.

FERRIC METALS, based on iron, are the most common thing on this planet and the fourth most common in the crust. Mix in a touch of carbon (less than 2 percent) and you have steel; more and you have cast iron.

ALUMINUM IS COMMON, SOFT, AND LIGHTWEIGHT. EASY TO MACHINE, FORM, AND CAST, BUT DIFFICULT TO WELD. IT'S A SOUGHT-AFTER SCRAP MATERIAL, SO SCAVENGING MIGHT BE SKETCHY.

IS IT IRON? TAKE A GRINDER TO IT. SEE SPARKS? THOSE ARE BITS OF IRON BURNING. FERRIC COMPOUNDS WILL SPARK, NON-FERROUS WILL NOT. OTHER TESTS INCLUDE EXPOSING IT TO OXYGEN AND STICKING A MAGNET TO IT, BUT SPARKING IS THE ULTIMATE WAY TO TELL.

A TINY CHANGE IN COMPOSITION CAN HAVE SIGNIFICANT CONSEQUENCES. A 1 PERCENT CHANGE IN CARBON IS THE DIFFERENCE BETWEEN BRITTLE CAST IRON AND THE HAMMER THAT SMASHES IT.

METALS ARE MALLEABLE. THEY HAVE A UNIFORM CRYSTALLINE STRUCTURE, WHICH ALLOWS THEM TO BE STRETCHED, TWISTED, MELTED, AND REFORMED AND STILL STAY METAL. METALS' INHERENT INSTABILITY—THEIR WILLINGNESS TO LOSE THEIR OUTER-SHELL ELECTRONS—RESULTS IN A FREE-FLOWING ELECTRON CLOUD THAT EASES ENERGY TRANSFER. THE RESULT IS HIGH CONDUCTIVITY OF ELECTRICITY AND HEAT.

THE METHOD USED TO WORK STEEL CAN EFFECT DEEP CHANGES WITHOUT YOUR KNOWING IT. DRILL A HOLE IN STAINLESS STEEL TOO SLOWLY, AND THE FORCE OF THE BIT WILL HARDEN THE METAL IN AN INSTANT, SNAPPING OFF THE BIT AS IT TRIES TO CUT MATERIAL SUDDENLY HARDER THAN ITSELF.

CONNECT DIFFERENT METALS AND YOU WILL CREATE ELECTRICITY. GOOD IF YOU'RE BUILDING A BATTERY; DISASTROUS FOR ANYTHING ELSE. ALWAYS INSULATE YOUR METALS.

105 UNDERSTAND ELECTRICAL WELDING

Lots of maker types start out with nonpowered oxy-acetylene welding (see #46–54), but fusing metal with electricity is inarguably more badass. At the core, all three common types of electrical welding—stick, also called arc welding; MIG (which stands for metal inert gas); and TIG (tungsten inert gas)—are really all the same.

Electrical welding creates heat via an electrical short circuit. The welding machine converts line power (110-volt or 220-volt AC) to low voltage (~30 volts), high-amperage (20–300 amps) electrical power—usually DC, but AC for some specific processes. One electrical line travels from the machine to the pieces to be welded (the "ground"—it can connect directly to the work or to a surface that conducts to the work, like the welding table). The other line makes its way to the electrode and arcs across to the work, completing the circuit and turning the electrical energy into heat that melts the metal, creating a weld.

A focused stream of inert gas protects the weld from corrupting oxygen and stabilizes the arc. This gas is pumped in from a tank with TIG and MIG welding or created by vaporizing flux in stick welding. In MIG and stick welding, the electrode is consumed and deposited into the weld as filler metal; in TIG, the tungsten electrode

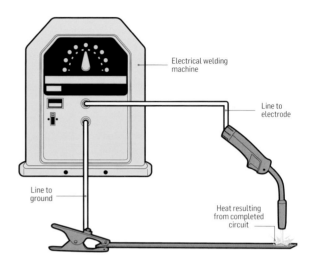

remains and filler metal is added by hand.

Electrical welding rigs range in price from free (such as TIG machines hacked together from microwave oven transformers) to affordable (low-end, used stick and MIG welders are easy to find at US$100–$300), all the way up to the new-car price tags on high-end TIG or multifunction machines. Choose the machine you can afford and that fits your needs.

ELECTRICAL WELDING PROS AND CONS

MIG
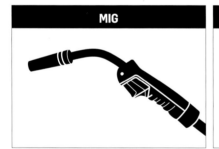

PROS Like using a hot-glue gun for metal—so easy a kid at summer camp could do it. This variety is easiest to learn, quickest to do, and moderately clean. Works with or without gas.

CONS MIG welding is expensive to get started in and it works on a limited range of metal types and thicknesses. It's not particularly clean or fine, and doing it gasless creates outright messy results.

STICK
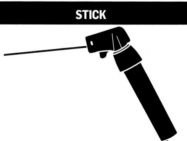

PROS The default if someone says "welding," stick has the longest history, is cheapest for beginners, is good for heavy and thick stuff, and is usable anywhere, anytime. Plus, different rods work on a wide range of metals. It can cut, too, and the equipment can really take a beating.

CONS It's messy, slow, and creates a lot of slag. Stick welding on thin materials is difficult to impossible and it's hard to do fine work.

TIG
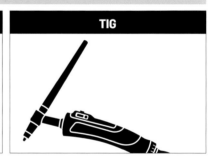

PROS This is the cleanest, most precise welding method out there—it also works on a lot of materials. When you're TIG welding, you can work at a very wide range of depth and penetration. TIG machines often include a stick-welding function.

CONS It's the most expensive and the hardest to learn (control, control, control). Go easy on the delicate equipment; know that gas is mandatory.

106 SET UP A SPACE FOR ELECTRICAL WELDING

Go back and read #46 on kitting out a station for powerless oxyacetylene welding. You need all the work-space parameters discussed there, plus this crucial stuff:

GET POWER Electrical welding requires electricity, and a lot of it. The smallest 110-volt welding machines need a dedicated 20-amp circuit, but 30 to 50 amps is better. Better yet is to run a dedicated 220-volt circuit with at least 30 amps and use a 220-volt welder. (Some of the nicer MIG and TIG/stick machines will run off either 110 or 220).

GET A WELDER If you're new to welding, broke, or both, you can make fine welds with three car batteries and jumper cables (see #111–112). If you have some money and just want to weld occasionally, get a MIG setup (make sure it can run flux core for gasless welding). If you're going to be doing a lot of metalwork or aspire to welding as a trade, get a TIG/stick machine—the portable AC/DC 100/220 machines are a viable business in a carrying case. You can always start small and upgrade as interest and funds grow.

GET SOME GAS TIG and MIG welders require tanks of inert gasses from your local welding supplier or gas vendor. To start, MIG welders will need a mix of argon and CO_2; TIG welders need pure argon.

GET SAFE Electrical welding emits a ton of blinding, health-harming UV light. You need a full-face helmet, with lenses and extra lens covers. Cover all your skin—even the darkest skin can be burned by welding light. Gloves, safety glasses, and a welding jacket or bib are also a must. **Pro tip:** Welding gets hot fast, and sweaty clothing can conduct the charge to your heart instead of the work. Keep dry and wear only nonsynthetic fibers so clothes don't melt and stick to your skin.

GET CONSUMABLES Each welding process uses specific supplies. MIG welders require rolls of wire and torch tips, TIG machines electrodes and filler rods, and stick welders stick electrodes. Have the stuff you need.

GET TOOLS Wire cutters and mechanics' pliers are helpful for cutting TIG rods and MIG wire. You could also stand to have a chipping hammer for stick welding.

GET CHEATERS Assemble a collection of pieces to clamp to, and use as jigs while you're welding. Metal expands and contracts as it heats up, but a tight clamp on a solid mass keeps the wiggle down. Also, some cheaters (such as bricks, scrap aluminum, or steel) will absorb excess heat that can cause warping.

107 START WITH STICK WELDING

To get you acquainted with the many tongue-blistering flavors of electrical welding, let's walk through one common weld—an inside corner—using stick, TIG, and MIG methods. Stick welding requires the least amount of initial investment.

STEP 1 Use a 6011 or a 6013 rod ⅛ inch (3.2 mm) thick or smaller. Set the welding machine to around 100 amps.

STEP 2 Position yourself and do a few test gestures. Make sure your arm can go through the whole range of the weld unobstructed.

STEP 3 Grip the electrode holder in your dominant hand. The rod should be pointing directly at the seam where the two pieces meet, at a 45-degree angle, about ¼ inch (6.5 mm) above the metal. Strike an arc and tack one end, maintaining the 45-degree angle. Go to the other end and do it again. Clean the tacks with a chip hammer and wire brush.

STEP 4 Weld the seam. Start on top of the first tack. Strike the arc, hold it for a couple of moments until you see the glossy puddle, then dip the rod into the melt. Envision the tip diving ⅛ inch (3.2 mm) into the puddle, then coming back out. Make a string of consistent, half-overlapping puddles all the way down the seam.

STEP 5 Clean your weld with a wire brush.

108 MIG WELD AN INSIDE CORNER

Where stick welding uses coated rods that need to be replaced as they are consumed, MIG welding torches can continuously pump out wire. MIG welding creates cleaner welds than stick welding, but it requires a bit more equipment and a touch more finesse.

STEP 1 Tack the ends. Position the tip of the MIG gun at one end of the work, wire extending ⅜ inch (9.5 mm) from the tip. The wire should be just touching the seam; the tip should be at a 45-degree angle, bisecting the corner. Pull the trigger and position a dot of weld equally distributed on both pieces. Repeat on the other end of the weld.

STEP 2 Position the cone so that its end is ½ inch (1.25 cm) past the tip. Jam the cone into the corner so that it makes a hypotenuse for the two pieces of metal. Point the wire directly at the seam at a 45-degree angle vertically. Lean the gun back so that the tip is at a 45-degree angle horizontally, the face of the cone balanced on the two pieces of steel.

STEP 3 Pull the trigger. Wiggle the tip of the gun in small, smooth, regular arcs, 90 degrees to 45 degrees to -90 degrees, all while pulling the gun toward the other end of the weld. Take your time and keep your eyes on the weld, making sure the deposit is uniform.

STEP 4 When you reach the last tack, overlap it completely and release the trigger.

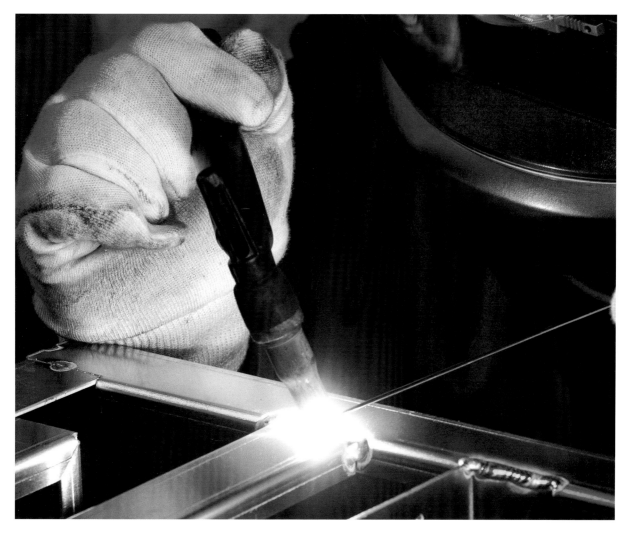

109 TRY SOME TIG WELDING, TOO

Of the three welding processes here, TIG welding is the most precise and most difficult to master. A TIG welding system allows you to control the intensity of your arc with a foot pedal—not unlike hitting the gas in a car. Also, TIG electrodes aren't consumed; welds are filled with a rod.

STEP 1 With the welder off, loosen the back cap until the tungsten slides freely. Hold the gun so that the nozzle face provides a hypotenuse to the two pieces of steel; the tungsten should be at a 45-degree angle, pointed directly at the seam. Position the tungsten so that the tip is about ¼ inch (6.5 mm) from the seam. Tighten the back cap.

STEP 2 Set the heat. A good rule of thumb for TIG welding steel is 1 amp per $1/1000$ inch (0.03 mm) at the top end, taking off 20 percent after the puddle has formed. Since $1/8$ inch (3.2 mm) equals 0.125, set the machine to 125 amps.

STEP 3 Holding your head so that you are looking straight down the length of the weld, get your rod ready. You should be using $1/16$-inch (1.6-mm) steel filler rod that it is long enough to get to the work's far end.

STEP 4 Position the gun in the 45-degree straddle at one end of the seam. Push the pedal all the way down; the arc should jump directly to the seam. As soon as the seam puddles, let up on the pedal. Then tack the other end.

STEP 5 Move the gun back to the first tack. Floor the pedal. As soon as you get a full, glossy puddle, back off 20 percent. Slide the rod down the seam and dip it in the puddle, pivoting the gun away to avoid crossing rod and arc. Move the puddle, add more rod, and repeat until you hit the last tack. Ease up on the puddle slowly. You should have a pretty weld that looks like it was made by a robot.

HACKETT SAYS

110 WELD LIKE A BADASS

Welders love the aura of instant fierceness that comes with putting on a welding helmet—we cultivate it, but we know that it is mostly unjustified. In reality, all crafts can be dangerous: Sewing needles are sharp, hammers love thumbs, and when the glue-gun label states that "The tip gets very hot," it sure does mean it. With these crafts, the danger starts at zero and then scales with more advanced use.

But with welding, the scale starts a lot higher. Even the most casual of toe-dipping (watching someone weld) requires special equipment. Watch someone shape wood with a bandsaw and you get to see the technique—the worst thing you risk is a little irritating sawdust. Watch someone weld, and you get a vague, dazzled idea of what they are doing, followed by the deep, unique pain of retinal burns and (usually) temporary blindness. Get a little closer and the spray of sparks can nail you.

Welders learn quickly to be prepared and to work within the envelope of risk. But it's worth it, because welders do things that are impossible in other maker disciplines. For example: At the core, welding is profoundly different from all other types of joining. Solder, bolt, glue, sew—pick a process, any process, and what you are doing is taking three items (the two things being joined and the joining agent, such as thread, solder, or hardware) and linking them together. The link could be at the molecular level or the arm-size-bolt level, but the key thing is that three items entered, three closer items leave. But with welding, you put in three items (stuff to be welded together, filler metal, and a rod, wire, or stick), and you get one item out. At the most basic level the things are now one—in fact, a common inspection process for critical welds is to x-ray the joint. Inspectors look deeply into the internal structure of the weld—if they can see where one thing starts and another ends, the welder did it wrong.

Welding can be hard. It can take a long time to get good at it. It can be dangerous. These things are true of anything worth knowing. Take care when setting up your work space, keep some basic rules in mind while working, and know the risks to those around you, and you will see welding is no more dangerous than a glue gun, less of a risk to your life than a table saw.

111 HACK TOGETHER A WELDER

Someday, when the end of the world as we know it comes, I like to think that I will be ready and willing to rebuild civilization from a pile of trash. When the grid goes down, all you need is a pocketful of tools—and a few car batteries.

STEP 1 Using jumper cables, connect three car batteries in series—the positive of one battery goes to the negative of the other, right down the line, until you have one loose negative and one loose positive lead. **Pro tip:** DC welding requires high-amperage, relatively low-voltage electricity. Wire the batteries in series and the result is 36 volts at whatever amperage the battery can hold. The more amp hours (usually, the larger the battery, the higher the hours), the longer you can weld.

STEP 2 Attach the loose positive lead to the work, as the ground for the circuit.

STEP 3 Since securely gripping a welding rod in jumper cable clips can be awkward, take a moment and bend back a couple of the cable clip's teeth until you have a secure spot and a lot of surface contact for the welding rod. You might want to wrap an inner tube around the handles to increase the power of the spring, or use a vise grip to hold the rod in place. A high-amp current is going through this connection, so make sure it is a good one.

STEP 4 Attach the negative lead to the welding rod. And there—you have a car battery welder.

STEP 5 Keep the leads isolated until it's time to put it in action. The leads are ready to weld, even if you are not. Plus, make sure the electrode is not touching the ground cable or anything that is conducting to the ground cable, like the work or a metal welding table.

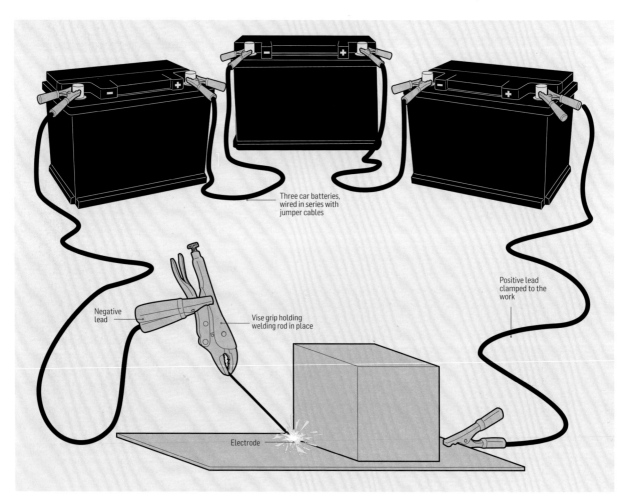

112 FUSE METAL WITH YOUR DIY WELDER

Welding is shrouded in an aura of mystery and menace, most of which comes from the high barrier to entry. Since you've built a welder, that's no excuse. Spend US$20 on a cheap helmet and another US$5 on 1/8-inch (3.2-mm) welding rod—6011 is a good, all-purpose rod that's available worldwide. Find some scrap steel (if it rusts, it's steel), put on some clothing that covers you completely and that you do not mind ruining, and get to welding.

STEP 1 Secure your work using clamps or a vise. The important thing is that all parts you want to weld together are conducting to one another and to the ground clamp.

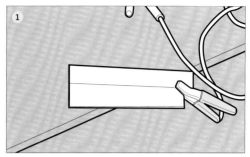

STEP 2 Strike an arc. Hold the welding rod holder in your dominant (and safety-gloved) hand. Firmly sweep the tip of the electrode against the work. Like a match, it should flare up—also like a match, it could take multiple tries to spark. Practice until you can get the tip of the electrode to stop about 1/8 inch (3.2 mm) from the work. The arc should be maintained, melting the electrode and the metal beneath it. **Pro tip:** If your welding rod gets stuck, break the circuit by releasing the electrode from the clamp. When it cools a little, pop it off with pliers or a hammer. Toss the electrode, or use it as filler rod.

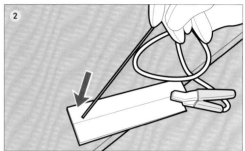

STEP 3 With the arc struck, hover for a moment where you want to start your weld. You should see a puddle of molten metal at the focus of the arc. This will be your first tack weld. In general (and as we've gone over in other welding sections), you should put in a tack at the beginning and end of where you will be running a weld, and every 6 inches (15 cm) or so in between.

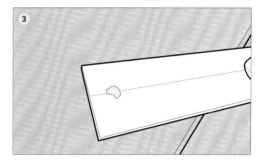

STEP 4 With tacks in place, strike an arc, then make a puddle of molten metal. Push the rod in to fill the melt pool, then move it forward a little (about half the diameter of the weld pool) and repeat. Imagine that you are making a red-hot stack of overlapping dimes. Experiment with it to get good.

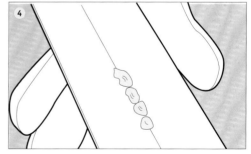

STEP 5 Your weld will be covered with a weird, glassy substance. This is slag—a protective layer formed by the flux on the outside of the rod boiling away. You need to clean the slag off to see your weld. Whack at the slag with a hammer. (While any hammer will do, a welding chip hammer is ideal.) Slag is brittle stuff, so even a glancing blow will break it off—make sure you're wearing eye protection.

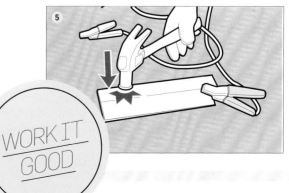

WORK IT GOOD

113 SET UP A SOLDERING STATION

Tinkering with electronics is one of the quiet sinkholes of joy and obsession that line the maker landscape. Fix a plug or make an LED blink, and the next thing you know 30 years have passed. That all starts here.

STEP 1 Choose a location. If you have a single worktable, devote a corner of it to electronics work, and run a dedicated power strip there. Soldering does not play well with vibration, metal chips, sawdust, or wetness, so plan accordingly. Ventilate by using a fume extractor (simple DIY versions abound). If you don't have one, mount a desktop fan to pull fumes away.

STEP 2 Get a soldering iron—a 50-watt, fine-tip one will do the trick for most projects. **Pro tip:** Wire a plug into a dimmer switch. Plug in the soldering iron, and its temperature is now adjustable.

STEP 3 Set up a soldering stand and third hand. A soldering stand holds the iron and cleaning sponge, while a third hand—basically alligator clips mounted on an armature with a solid base—can hold materials as you solder. **Pro tip:** Actual sponges cool the iron and inevitably char. Use a copper-pot scrubber instead.

STEP 4 Stock up on parts and tools. Desoldering braid, small screwdrivers, sidecutters, tweezers, and solder suckers all come in handy. There are too many little electronics components to list, but as you move from project to project, always get more than you need.

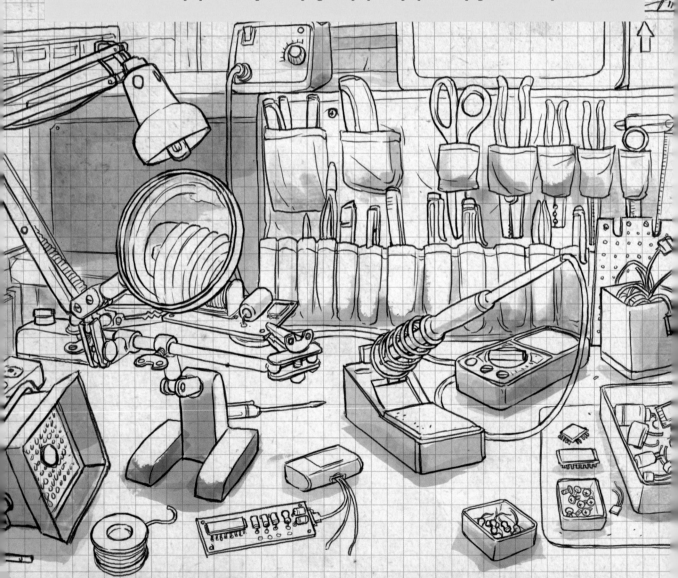

114 SOLDER TO A PCB

Printed circuit boards (PCBs) are the standard for electronics making. Whether you get them in a kit, scratch-build on plain ones, or design and print your own, most small-scale electronics can be reduced to through-hole soldering.

STEP 1 Secure your board in a vise. **Pro tip:** Secure the PCB vertically. Getting lost on a PCB is easy; securing it vertically lets you compare sides instantly.

STEP 2 Insert components according to your diagram or breadboard, with the components on the naked side, and their leads emerging through the copper side. It's easiest to start at one end and work straight through.

STEP 3 Secure components by sharply bending the leads to the copper side of the board. **Pro tip:** Excess lead material can be used as traces; just bend them to run wherever you need them.

STEP 4 Touch the tip of the soldering iron to the crux where the copper and lead meet. Wait a beat, then touch the tip of the solder to the other side of the lead. It should melt and fill the gap. Repeat for all connections, then snip off excess lead wire.

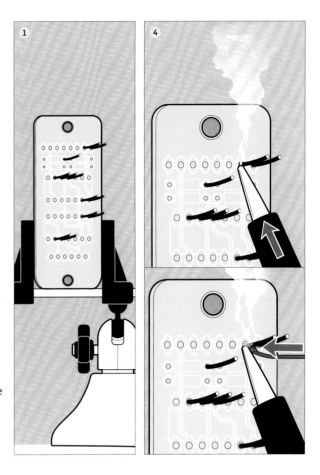

115 SOLDER WIRES TOGETHER

Fix a plug severed by an errant circular saw, put a switch where you want it, or shoot off fireworks from the comfort of your couch. Soldering wires together is a core electronics task.

STEP 1 Strip the wires, removing between ¼ to ½ inch (0.65–1.25 cm) of insulation from each end. (**Pro tip:** Standard strippers have little notches marked for different wire gauges, but the radius gap between your incisor and canine makes a perfect stripper for 18- to 22-gauge wire. Just bite, pull, and spit.) Hold the two wires between thumb and forefinger of your nondominant hand, fingertips pinching the insulation, wires at a right angle. Grab the V they create with the fingertips of your dominant hand and twist the wires together.

STEP 2 Jam the side of a hot soldering iron on one side of the pigtail. Wait a beat as the wire heats up, then touch the solder to the wire opposite the iron.

STEP 3 Push solder until you can see it seeping out on the iron side.

STEP 4 Let the joint cool, then wrap it in vinyl electrical tape.

116 DECODE CIRCUITRY SCHEMATICS

A schematic is a map of a circuit. Like a good map, it tells you exactly where things go—and exactly how to get to where you want to be. You can be amazingly, blindingly ignorant of electronic theory; parallel resistor math can enrage you like a bird in a mirror. You do not even need to know what the circuit does. But if you learn how to read schematics, the resulting circuits will be masterful, and everyone will be fooled into thinking you know what you're doing. And then, someday soon, you will.
Pro tip: Datasheets are an invaluable source of reference schematics and advice.

Schematics have two parts: symbols that represent electronic components, and lines that show the connections between said components.

COMPONENTS Components get a symbol, plus a letter and number designator. The letter indicates the type of component: R for resistor, C for capacitor, and so on. Tricky ones are Q for transistor and IC for chip. The number is to separate the different types of that component in the schematic. Somewhere on the schematic will be a parts list outlining different component values. This will tell you the exact parts needed to build the circuit.

COMMON COMPONENT SYMBOLS

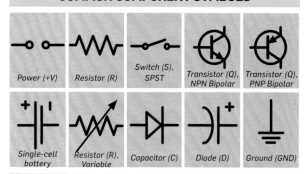

CONNECTIONS Schematic lines between two components just show that they are connected, not how they are connected. Could be wire, could be a copper trace—if it is important, it will be specified somewhere else in the documentation. Otherwise, any conductor that fits will do.

TYPES OF CONNECTIONS

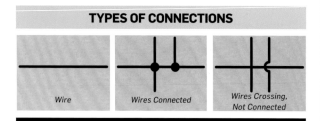

117 PROTOTYPE WITH A BREADBOARD

A schematic shows you the logical connections of a circuit's components, but it doesn't ensure the circuit will work. To prototype a design and tinker until you get your connections just right, it helps to use a breadboard. This fundamental piece of equipment is great for both beginners and circuit Jedis, as it can house simple and complex circuits alike.

CHECK OUT THE GUTS Open a breadboard and you'll find several horizontal rows of metal strips—these metal rows are conductive and will allow current to flow. At the tops of the rows are tiny spring-loaded clips (called contact points) that let you stick a wire or a component's lead into an exposed hole to secure it in place.

MIND THE RAILS On the outside edges are the rails: rows of positively or negatively charged holes connected horizontally so that power can flow where it is needed. You can connect a power source to a single point on a rail and have access to that power at any other tie point on that rail. Power rails are usually demarcated with red and blue lines.

AVOID SHORTING A rule of thumb is to not put all of your component leads into a single row of five tie points—since they're already electrically connected, you'll create a short circuit. Instead, connect one lead of a component to one row of tie points and the other lead to a different row to keep them electrically separated.

118 ETCH YOUR OWN CIRCUITS

Making your own circuit boards might sound like far-out wizardry, but they're relatively simple and inexpensive to make. You can get your hands on a circuit board for a few bucks, transfer your circuit design onto it, and then use chemicals to etch into and reveal the conductive copper. You'll be using chemicals that can damage your skin or even blind you, so use proper safety gear (goggles and gloves are a must), read the instructions on your supplies, and always work in a ventilated area.

STEP 1 Design your circuit, either sketching it or creating it in software (which will likely tell you if your circuit will work before you print it; otherwise, use a breadboard to test your hand-drawn circuits). Print or draw the circuit onto thin, cheap paper.

STEP 2 Wearing safety goggles and gloves, and working in a ventilated area, gently pour 1 part hydrochloric acid for every 2 parts hydrogen peroxide into a nonmetallic basin that can withstand acid.

STEP 3 Press the paper to your exposed copper plate to transfer the pattern, then submerge the board completely into the solution and agitate it for 10 to 15 minutes. (Don't put your face directly over the solution, because it will eventually get warmer as you agitate your board and start to fume even more.) Continue to stir until all the copper (except for the copper under your lines) has dissolved, and the solution has become slightly green.

STEP 4 Rinse the circuit board using cold water and wipe it down with a rag or paper towel to remove any remaining etching solution. Then, mix a 1:1 ratio of acetone and rubbing alcohol and, using a paper towel, gently rub the surface of the board until the markings come off completely, revealing the lines of conductive copper underneath.

STEP 5 Now it's time to drill holes for your components' leads. Before you start, locate and mark all the positions. Wear a dust mask as you drill, as copper dust is toxic stuff.

STEP 6 Place the components onto the circuit board at their designated locations. Be careful that parts with polarity are lined up correctly with the corresponding positive and negative ends. Test to make sure it works before soldering the components in place.

STEP 7 Finally, etching is toxic not only to us but to other creatures as well. When you're done with your etching solution, dispose of it by labeling it and handing it off to a chemical-waste facility.

MAKE YOUR OWN

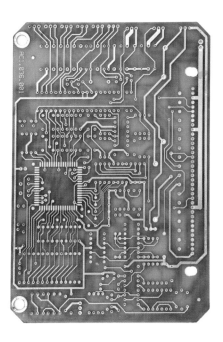

"Only suckers pay a premium for custom printed circuit board. Plus, you get to make an acid bath without landing on a terrorist watch list."

119 PICK ELECTRONIC COMPONENTS

Components are the building blocks of any electronics project. They usually have two or more electrical leads and when they connect, typically with the help of a printed circuit board, technology happens.

RESISTOR The opposite of a conductor, basically anything that electricity has a tough time moving through. Like a kink in a hose that slows down the amount of water that can flow through, a resistor slows down the amount of electricity flowing through a circuit.

WIRE Solid, stranded, or braided, all made from a handful of particularly ductile, conductive, and strong metals and alloys. One of the most notable—and most often stolen for scrap—is copper. For the sake of soldering, wire used inside many gadgets is often tin-plated.

SWITCH At its most basic, simply a component that interrupts or redirects an electrical current. From the standard light switch to your doorbell, to the knife switch that brought Frankenstein to life, switches all serve to open and close circuits.

INTEGRATED CIRCUIT Where traditional circuits are made up of separate components attached to a shared substrate, integrated circuits—a.k.a. chips—combine those components onto a single plate for easy use.

CAPACITOR Essentially two metal plates separated by a nonconductive substance, made to store electricity. Careful, though: They can dump their whole charge in one go. Touch the wrong one and you're toast.

TRANSISTOR If a resistor functions like kinks in a hose, then a transistor is the valve that controls the flow of water. Capable of turning current within a circuit on and off or amplifying a current. You'll find these inside everything from old radios to computer chips.

STEPPER MOTOR Converts electrical power into mechanical energy but can be told to stop at certain steps. Controlled by switching on and off a series of electromagnets around an often toothed, gear-shaped center rotor. Common in computer printers.

SENSOR Takes a physical stimulus and converts it into a signal that can be understood by any number of people or gadgets. They're the reason you can go into a public restroom, flush the toilet, then wash and dry your hands, all without touching anything.

TRANSFORMER Allows for energy transfer between circuits, as well as changes in voltage. Step-up transformers increase voltage, while step-down transformers—like the one inside your phone charger—decrease voltage, helping your phone charge instead of melt.

COIL/CHOKE At its most basic, consists of a wire coil wound around a core. In a lot of electronics, one or more chokes usually prevent alternating current from reaching certain parts of a circuit—"choking" it—while allowing DC through.

POTENTIOMETER This component essentially functions as a resistor that can be adjusted manually with a knob, slider, or other control mechanism. If you've ever changed the volume on a stereo, or adjusted the speed of a fan, you've used a potentiometer.

120 KNOW YOUR LEDS

In my life, I have encountered roughly a million billion LEDs and exactly zero explanations as to how they work. Essentially, a charge moves across a semiconductor; a semiconductor has a "high" side and a "low" side, one holding more energy than the other. When current flows down from the high side, electrons drop to a lower energy level to fit on the low side. That drop emits photons, the building blocks of light. Here is some other useful LED information:

VOLTAGE MATTERS Too small a voltage, and the LED will do nothing. Too high, and it'll get really bright and then crap out on you.

POLARITY MATTERS If you wire an LED in backward, the diode will block the circuit. **Note:** If your LED is new, the longer wire is the positive one. On used or old-school LEDs, feel around the rim of the bulb. There should be a flat spot or notch on the negative side.

TOO MUCH CURRENT KILLS While an LED will take all of the power it's given, unlimited amps will kill it. Use a resistor to meter out current.

WHAT YOU SEE IS WHAT YOU GET An LED's color and brightness is set when it's built. If you want color or brightness changes, you have to cheat by adding additional LEDs or tricking the eye with strobing. Or use a multicolor LED to begin with.

121 WIRE UP THE SIMPLEST CIRCUIT

Wiring an LED is pretty much the simplest possible circuit. Since coin cells put out a limited amount of current, this is basically the only instance where you don't need to wire in a resistor. Slide the coin-cell battery between the legs of an LED so the LED's positive (slightly longer) leg touches the positive side of a 1.5-volt coin-cell battery and the LED's negative (shorter) leg touches the battery's negative side. Secure the LED into place with electrical tape.

122 STORE ENERGY WITH A CAPACITOR

A capacitor (a.k.a. cap) is a two-lead electrical device that responds to direct current (DC) and alternating current (AC) in very different ways. In a DC circuit, a cap often functions as an energy-storage element. Connecting a battery across the leads of a capacitor will cause it to charge up. Disconnect the battery, and the energy will remain stored for a while (forever, in an ideal capacitor, but of course we don't live in an ideal world). Touch, say, an LED to the leads of the charged capacitor, and the LED will glow briefly as the capacitor discharges through it. "Larger" capacitors take longer to charge up, and this effect makes them useful for introducing time delays into a circuit, as well.

DC cannot pass through a capacitor, but AC can—though the efficiency depends on the frequency of the AC in question and on the size of the capacitor. In an AC

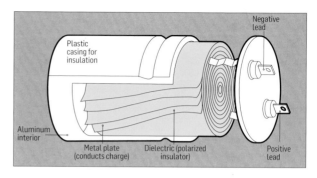

circuit, a capacitor often functions as a "filter" element. One common use, especially in audio electronics, is to filter out DC noise from a circuit and leave only the AC signal containing the audio information behind. Unlike resistors, connecting capacitors in series doesn't cause their values to add up. If you want to substitute several smaller caps for a bigger one, you have to connect their leads in parallel.

123 REGULATE WITH A POTENTIOMETER

A potentiometer (a.k.a. pot) is a three-lead electrical device with a rotating knob. It may be large and prominent, for easy manual adjustment, or it may be small and hidden inside a case, with a knob designed to be tweaked occasionally with a screwdriver. The small, hidden kind is sometimes called a trimmer.

A pot's middle lead is electrically connected to a rotating wiper inside. The wiper slides along a length of wire coil or other resistive material, which is in turn electrically connected at its ends to the pot's two outside leads. The wiper moves when you rotate the knob, causing the resistance between the middle lead and the two outside leads to change. Turning it one way causes the resistance at the left lead to go up while the resistance at the right lead goes down, and turning it the other causes the opposite effect.

The potentiometer's middle lead must be connected for the knob to have any effect, but for its most basic use as a variable resistor (a.k.a. rheostat), only one of the two outside leads must be. Which you choose doesn't matter much electrically, but it will affect the handedness of the knob's rotation (i.e., whether turning it clockwise introduces more or less resistance into the circuit). It's common practice when wiring a pot this way to short the unused outside lead to the middle/wiper lead. This will prevent the open terminal from becoming an accidental antenna and introducing noise into the circuit.

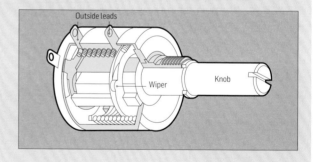

"Where there's current, there's desire to control it—to nudge it up or down, redirect it, or otherwise show it who's boss. These simple tools help you do just that."

124 DIRECT CURRENT WITH A RELAY

A relay is a mechanical switch operated by an electrical current. It has at least four leads—two for the current that is being switched and two more for the current that is doing the switching—but will often have many more. As with manually operated switches, relays come in a large variety of configurations in terms of the number of circuits that are switched when the relay operates ("poles") and the number of positions through which those circuits can be cycled besides open ("throws").

Thus there are single-pole single-throw (SPST) relays that switch a single circuit on or off, double-pole single-throw (DPST) relays that switch two circuits on or off, single-pole double-throw (SPDT) relays that switch a single circuit between two possible output paths, etc. As with switches, relays can be "momentary" (requiring continuous energy input to remain in their "active" state) or "latching" (switching from one state to another, and staying there, each time energy is applied).

Very often, relays are used to switch a high-power device (like a motor) using a low-power one (like a timer or a microcontroller). The output from a microcontroller could never power a large motor directly, but it can switch on the flow of current from another source, like a car battery or wall outlet, using an appropriate relay.

Relays have existed since the early 1800s, and though there are now modern semiconductor devices (like transistors and so-called "solid state relays") that can serve most of the same functions, good old-fashioned electromechanical relays have advantages (like a large variety of pole/throw combinations) that will likely keep them in widespread use for a while to come.

125 CALCULATE LED RESISTOR NEEDS

Most LED circuits will require a resistor to keep instant death at bay. Ohm's law—Resistance = Voltage/Current—tells us what we need. Let's run a typical red LED, which probably draws 20 milliamps at 2 volts off a 9-volt battery. The voltage we need to resist is what the battery can give (9 volts), minus what the LED needs (2 volts). The bulb will need 20 milliamps (.020 amps) so:

$R = (9-2)/.020$
$R = 350\ ohms$

BUILD

STEP 1 Cut two lengths of wire, one 2 inches (5 cm) long, the other 4 inches (10 cm) long. Strip ½ inch (1.25 cm) of insulation from each end.

STEP 2 Attach the 4-inch (10-cm) wire to the negative (short) leg of the LED, and the 2-inch (5-cm) wire to the positive leg. Twisting them together will work; solder if you feel fancy.

STEP 3 Attach the positive wire to a 350-ohm resistor. Again, twisting is sufficient.

STEP 4 Attach the negative wire to the negative battery terminal. Secure with tape.

STEP 5 Touch the free end of the resistor to the positive lead of the battery. Let there be light. Secure with tape.

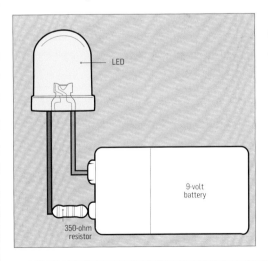

126 PICK ILLUMINATION SOURCES

No modern mad scientist ever toiled into the wee hours without some artificial light source illuminating his or her work. Lights also often play a role in the maker projects themselves—as indicators, info-flashing displays, or show-offy decorations. Here are some to choose from.

INCANDESCENT Electric current passes through a metal filament inside a bulb filled with inert gas, heating the filament until it glows. Only a small fraction of energy consumed is transformed into illumination, so it produces more heat than light.

FLASHING LED Sure, you could wire a special circuit to cause an LED to flash at certain intervals, but this one essentially does that job for you without the external circuitry. Equipped with its own multivibrator circuit, it flashes at set intervals when connected to power.

LED STRIPS LED strips are basically a series of bright LEDs—they're the ones you hope pretty much never turn on as the emergency lighting along the floor of many airplanes. Other, less terrifying uses include high-efficiency, easy-to-install cabinet and workshop lighting.

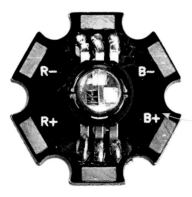

HIGH-POWER LED Pretty much exactly what it sounds like: Superbright LEDs that run at power tens and hundreds of times that of miniature LEDs. Since extra power creates extra heat as well as light, HPLEDs need a heatsink to draw heat away.

HALOGEN Functions like an incandescent bulb, with a type of halogen gas inside. Halogen gas and tungsten filament create a cycle by which some of the tungsten that evaporates from the application of current is redeposited in the filament—allowing for longer lamp life, higher temperatures, and brighter bulbs.

FLUORESCENT Remember those humming, occasionally flickering, bright-white tubes hanging above your school cafeteria table? Fluorescents are more expensive than many other sources of illumination because they use less energy, but it doesn't make them pretty.

MINIATURE LEDS Essentially tiny lightbulbs that last longer and lack a filament. They light up with the movement of electrons through semiconductor materials. Miniature LEDs represent the smallest end of the LED spectrum and can be useful when you're testing basic circuitry.

BICOLOR LEDS Really two LEDs in one. Current flows in one direction through the first, creating one color, and flows in the opposite direction through the second, creating another color. Can indicate the current's direction when you're designing circuits.

INFRARED Infrared lights use wavelengths outside the visible spectrum and are as ubiquitous as TV remotes and wireless mice. Infrareds have far less harmless applications in their ability to remotely gauge temperature, illuminate the darkness (as in night-vision goggles), or guide heat-seeking missiles.

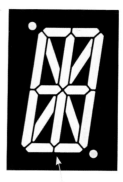

TRICOLOR/RGB LEDS Like their bicolor cousins, these have more than one LED in one case—each with a red, green, or blue emitter. Each of the LEDs can be controlled independently. Usually, each LED has its own lead, and a fourth lead is shared in common.

ALPHANUMERIC LEDS Can be programmed to display the full alphabet, numerical characters, and even some still graphics. Seen in roadside construction signs flashing instructions and the corny graphic display outside the local used-car dealership. Use them to slow local traffic at your will.

127 FADE AN LED

LEDs beat the pants off tungsten bulbs in pretty much every way. There's just one catch. If you want them to fade, you can't just drop the power—they're particular about how much juice they run on. In order to properly fade an LED, you'll need to do a little pulse-width modulation (PWM), a fancy phrase for blinking a light on and off too quickly for your eyes to tell. By flashing an LED more than 90 times per second and controlling the ratio of off to on, you can make an LED precisely as bright or dim as you like.

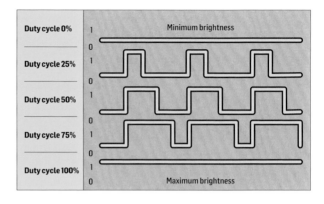

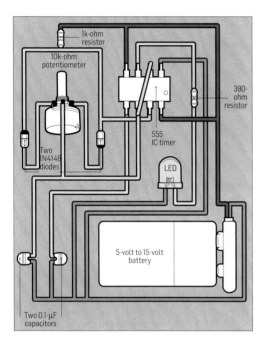

ANALOG METHOD

STEP 1 Referring to the diagram, wire the circuit on a breadboard.

STEP 2 Adjust the pot. You should be dimming the LED on pin 3. **Pro tip:** If you have a scope, you can monitor the output of pin 3. You'll see a pattern resembling the waveform in the above graph.

STEP 3 See the light go on and then off? Kudos. May your lights always be pleasantly flickering.

DIGITAL METHOD

Plug an LED into your Arduino (see #184) and try out the following:

```
void loop() {
  digitalWrite(led, HIGH);
  delay(1);
  digitalWrite(led, LOW);
  delay(10);
}
```

This is just the standard Arduino blink code with tweaked values. You're coding duty cycles by hand. Change the numbers, making sure they add up to 11 (11 milliseconds, or about 90 times a second). Neat, huh? Libraries are easier, but now you have an idea of what's going on.

128 WIRE IN SERIES OR IN PARALLEL

LEDs are a practical way to learn about parallel versus series wiring. Try powering two LEDs with a 3-volt battery by placing them in a row, with (+) on one end and (–) on the other. Nothing happens! Now take the same number of LEDs but place them side by side, so all the (+) leads are connected to the battery (+) and likewise with (–). See? It's fixed.

What's happening? A key rule in electronics is that components placed in series will require more voltage, while those placed in parallel will require more current. An average 5mm LED takes 2 or 3 volts at 20 milliamps. A 3-volt coin cell won't put out 6 volts without help, but on the other hand, it won't have trouble providing 80 milliamps.

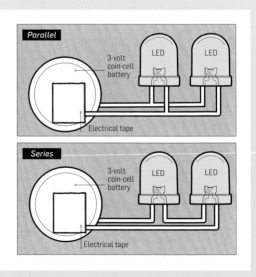

129 WORK WITH HIGH-POWER LEDS

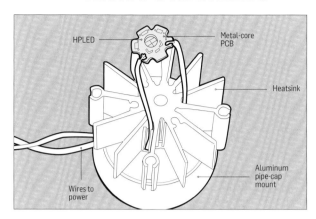

High-power LEDs (HPLEDs) are beasts. We're talking 0.5+ watts, not the 0.06-watt draw from a typical 5mm LED. These LEDs are perfect for projects where you need a ton of brightness—they pack an insane amount of light (up to 100-watt equivalent) in a tiny package. Simple current-limiting resistors won't cut it—you need drivers.

A current-regulating driver will be more efficient than a current-limiting resistor, which at high wattages will make a big difference. It'll run cooler and more reliably than a resistor. And many driver boards have pulse-width modulation (PWM) ability built in, to boot.

Try powering an HPLED without cooling and you'll burn your fingers and it. Efficient as they are, 70 percent of the power still comes out as heat. (**Note:** Keep LEDs under 212°F [100°C].) Many HPLEDs you'll find come mounted on a metal-core PCB (MCPCB), giving you something to mount a heatsink to. The MCPCB alone should have enough surface area to handle a 0.5-watt LED. Anything higher, and you'll need a heatsink to draw heat from the LED and dissipate it quickly. Good heatsinks maximize surface area, so look for hunks of aluminum with neat fins.

A 1-watt LED may need a heatsink about 2 inches (5 cm) wide. A 3-watt LED may want something 60mm–100mm square. For a 100-watt, think heat pipes, fans, and a massive brick of fins. But really, it's all about surface area. You can even use scrap aluminum instead of fancy sinks, as long as there's enough area.

STEP 1 Make sure the heatsink and MCPCB are as smooth as possible. You'll want to see yourself in them—even microscopic blemishes are bad. Remove every bit of dirt.

STEP 2 Add a bit of thermal grease to each surface. Its job is to fill in cracks too small for you to see, so only the thinnest layer is needed. Use plastic to spread it.

STEP 3 Mount together securely. Screws are good.

130 MASTER MULTIPLEXING

Let's say you have six free pins and need to control six LEDs. Sounds terrific, right? Now let's say you have six pins and need to control nine LEDs. Independently. Still terrific, because now you have a chance to learn about multiplexing.

Arrange the nine LEDs in a 3x3 grid. If you make one row high and one column low, the current will choose to flow only through the one LED specified. The magic happens when you cycle through all the rows really fast, taking advantage of the persistence-of-vision effect on the eye, so the flicker isn't noticed. As the grid grows, the pin savings add up.

STEP 1 Wire up a 3x3 grid of LEDs with resistors on one side. Connect the anodes together in each row and the cathodes together in each column.

STEP 2 Plug each row and column into a separate pin.

STEP 3 Using an Arduino or microcontroller of your choice, try some code, along the lines of:

```
digitalWrite(row[0], HIGH);
digitalWrite(col[2], LOW);
digitalWrite(row[0], LOW);
digitalWrite(col[2], HIGH);

digitalWrite(row[1], HIGH);
digitalWrite(col[1], LOW);
digitalWrite(row[1], LOW);
digitalWrite(col[1], HIGH);
```

Congratulations, you have now multiplexed. Go ahead, play around with it a while. This would be shorter in a practical application, but we're being educational. If you want to go further, look into things called "arrays."

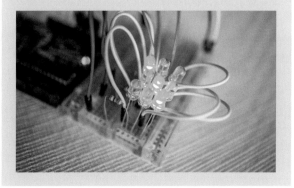

131 GET THE 411 ON THE 555

The 555 timer is not (quite) the simplest chip ever made, but it's the one most makers cut their teeth on. And it's crazy versatile: There are so many ways to use the 15 resistors, 25 transistors, and two diodes that make up the 555's guts that people are still dreaming up new ones, 40 years after it came out. Here are the 555's pins.

GROUND Every circuit needs one. The 555 is no exception.

TRIGGER Connecting this pin to ground turns on the output current and starts the timer.

OUTPUT Here's where you connect whatever you want to turn on and off—an LED, a buzzer, another 555, whatever.

RESET Connecting this pin to ground makes the timer stop and restart. Not often used in basic applications.

CONTROL Applying an external voltage here lets you adjust the default sensitivity of the trigger and threshold pins. It's not often used in basic applications, though it's common to ground it through a tiny capacitor to prevent electrical noise from causing weird behavior.

THRESHOLD Usually connected between a capacitor (which is like a tank that slowly fills with charge) and a resistor (which is like a valve that controls how fast charge flows into the tank). When the tank is full, the timer stops. Bigger capacitors take longer to "fill," and bigger resistors cause them to fill more slowly.

DISCHARGE Opens a connection to ground when the timer

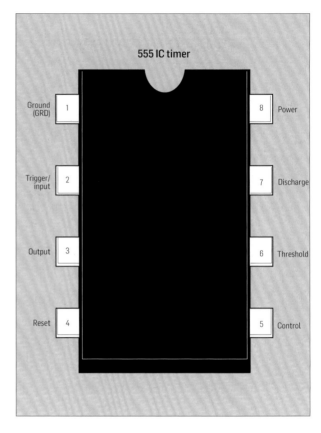

stops so that the capacitor can be emptied and made ready for a new interval to begin.

POWER Provides positive supply voltage for the 555 to run. Depending on make and model, this can be anything from 1.5- to 18-volt DC. Best to keep it between 4.5 volts and 15 volts until you really know your way around.

132 SET UP A FREE-RUNNING OSCILLATOR

Specifically a free-running flasher—no, I'm not talking about streaking. This dead-simple circuit flashes an LED on and off about once a second until the batteries run down.

STEP 1 Build the circuit on a breadboard. Here the 555 is wired in "free-running" or "astable" mode. The current from pin 3 turns on and off over and over again at a rate set by the resistors and capacitor connected to pins 7 and 6.

STEP 2 Speed things up by replacing the 10k-ohm resistor with a 1k-ohm resistor. The LED should now be blinking at a rate of about five times a second.

STEP 3 Speed things way up by swapping out the 100-μF capacitor for a 1-μF one. The LED will start blinking faster than your eye and brain can see (about 500 times a second), but not faster than you can hear.

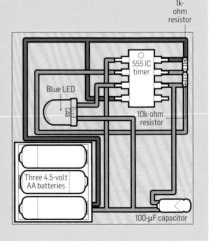

133 WIRE A ONE-SHOT TIMER

Replace the LED with a piezo speaker and your blinky becomes a buzzy. Now you have a most excellent circuit that will whine until it runs out of power or you smash it into bits. Try adding a second 555 wired in "one-shot" or "monostable" mode to turn the sound on and (mercifully) off again a moment later.

STEP 1 Add the new components and rearrange the jumper wires on your breadboard. See how the power input (pin 8) of the free-running 555 gets connected to the signal output (pin 3) of the new one?

STEP 2 Pinch the stripped end of the wire connected to the trigger pin with one hand, then briefly ground it by touching the brass backing of the piezo disc with the other. (Wait, let me take cover first. OK, go!)

STEP 3 Mess around with the values of the 10-μF capacitor and the 100k-ohm resistor to change the duration of the sound. Bigger capacitors and resistors will make the sound longer, and vice versa.

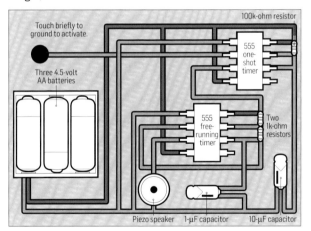

134 MEET THE 555 FAMILY

The 555 comes in 32 flavors and then some. Like most chips, it's produced in several packages ranging from round metal cans to tiny surface-mount devices with pads instead of pins. The original design, using TTL (transistor-transistor logic) technology, is still made and widely used, but a variant based on the more modern CMOS (complimentary metal oxide semiconductor) tech is also available and will run on a supply voltage as low as 1.5. Because many circuits (like the buzzer you just built) use more than one timer, multiple 555s are sometimes combined on a single chip. The 556 timer, for instance, contains two 555s in a 14-pin package, and the 558 has four timers spread across 16 pins.

"Crank too much power into an IC timer and it'll explode with a *POP* and spark that's perfect for igniting stuff. Harness that catastrophic failure by using one as a timed detonator."

135 GET TO KNOW YOUR MULTIMETER

In most areas of making, our senses tell us when something is broken or not quite right; the smell of smoke, a horrible grinding noise, or a sudden numbness where your finger used to be tend to give it away. Electronics are more subtle. If our senses pick up a problem, it probably means something is on fire. The multimeter is our window into the world of circuitry. It will tell you the voltage or resistance between two points, if two points are connected, and the amount of current passing through.

DIAL The big knob up front allows you to choose what characteristic you are measuring and the range to look in; the numbers (like 20 volts) show the top and bottom end of the range. If you choose the wrong range, the meter will display "ER" or "1." Just turn the knob and try again.

PROBES There should be two probes (usually one red, one black) and three holes: COM (common or ground = black probe), VΩ mA (voltage, ohms, milliamps = red probe), and 10ADC (for high-power checking; red goes here when checking current between 200 milliamperes and 10 amps).

LCD DISPLAY Spin the knob. Watch the decimal point move.

136 CHECK FOR CONTINUITY

Electronics are simple: Energy wants to go from a place of higher voltage to lower voltage along a circuit. When a circuit does not work, either the flow is interrupted, or the wrong amount of power is going to your components. Continuity is the first (and often last) thing to check when troubleshooting.

STEP 1 Turn the knob to "continuity." You may need to hold down the mode button to toggle between the continuity and voltage drop testers, which sometimes share a dial position, is indicated with a little diode symbol. No continuity setting? Select the lowest range of the resistance settings. Test by touching the two probes together. Some multimeters will beep, while others will give a reading close to zero.

STEP 2 Remove the battery or unplug the circuit; don't just flip off the switch.

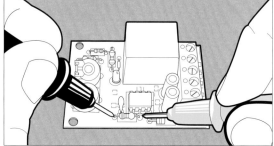

STEP 3 Simultaneously touch the two probes to the circuit, one to power and one to ground. If there's a gap, keep one probe at the beginning and step back along the circuit until you find the gap. If you can't find a gap, check power and ground to see if they're connected. If they are, you have a short in the system. Step back in the circuit until the problem disappears, then go forward until you establish the location of the short. **Note:** Switches are intended gaps. Set switches to "on" when troubleshooting.

137 MEASURE VOLTAGE

Problems are often caused by too little voltage where it's needed. Poor solder connections, wrong resistors, or parts drawing too much power are often to blame. Voltage measurements are done in parallel. You can piggyback into the wires while they're running and observe, silently judging like the NSA of the breadboard.

STEP 1 Set your multimeter to the correct voltage range. USB or battery-powered projects will probably be okay at 20 volts, electric vehicles or welding gear at 200, individual LEDs and components at 2. **Pro tip:** My multimeter has a 750 setting for AC volts. Actually checking something in this range will probably lead to a greasy stain where your body once was, as well as a damaged multimeter. If you need to check anything AC hotter than a power strip, use a no-contact multimeter.

STEP 2 Compare the system power to what it should be. If the system is all 5-volt components, make sure 5 volts are going in.

STEP 3 Check power drops. LEDs take a set amount of voltage—too much and the system stops dead. Also, a backward LED will cut the circuit.

STEP 4 Check voltage to the component's power-supply pin. If there's not enough voltage, most parts will not work.

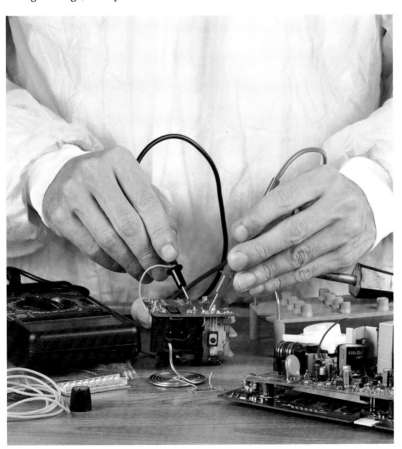

138 GAUGE POWER

Current is fundamental in electronics. While volts can be correct throughout a system, without sufficient amps there are no electrons to do the work. Measuring current can get tricky fast. A few things to keep in mind:

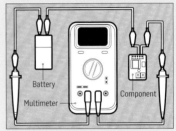

BE ONE WITH THE CIRCUIT
Current measurements must be done in series. The multimeter has to become part of the circuit. At the point where you want to measure current (such as between a resistor and an LED, or a switch and a motor), interrupt the circuit and swap in the multimeter like an ungainly wire. Power up the circuit; the multimeter will tell you how many amps are passing through.

CHOOSE YOUR RANGE WISELY
Most multimeters have a whole other port for above 200 milliamperes (mA; usually marked "10ADC"). There's no room for error; if you put a 500-mA load through the 200-mA setting, you'll learn all about multimeter fuse replacement. Failure will be immediate and catastrophic.

LOOK TWICE Always double-check your multimeter settings. Why? If the system is set on current (<200 mA) and you check the voltage, all of the system power dumps into the multimeter, frying it.

139 SEW CIRCUITS ON FABRIC

Most circuits and electronic components are rigid, inflexible, and can't easily be wrapped around curved or pliable surfaces, making them difficult to wear on the body. Soft circuits made from conductive fabrics and threads are an alternative solution to traditional printed circuit boards (PCBs) and wires—empowering you to create simple circuits that are flexible, washable, and move with you.

CONDUCTIVE FABRICS These are pretty much regular textiles, except they're plated or woven with metallic fibers—such as silver, nickel, copper, and stainless steel—that enable the textile to conduct electricity. They can be cut and sewn like traditional fabrics and made into flexible switches and sensors, ranging from touch sensors to textile electrodes that sense muscle activity.

CONDUCTIVE THREADS AND YARN These are composed of metallic elements such as silver or stainless steel and, similar to wires, can be used to create conductive paths for electricity to flow from one point to another. Unlike wires, they're flexible and can be hand or machine sewn, and can even be embroidered onto textiles to create sensors. An important difference between wires and conductive thread is that conductive thread is not insulated. To prevent a short circuit or the current from traveling along an unintended path, be sure your conductive traces don't unintentionally touch or overlap.

140 INSULATE YOUR CONDUCTIVE THREAD

When sewing by machine, alternate the conductive traces between the top and bottom of the fabric, your insulating layer. One advantage of using conductive thread in the bobbin and nonconductive thread as your top thread is that the conductive trace is isolated to the bottom of the fabric. You can alternate the location of the conductive thread by simply turning the fabric over when sewing your traces. Use bias tape or another insulating fabric layer to isolate your conductive traces.

141 PICK FABRIC FOR ITS RESISTANCE

Choosing the right fabric and thread for your project is a delicate balance between the physical and electrical properties of each material and the aesthetic and technical requirements of your project. The most important factor is the material's surface resistivity, which, measured in ohms (Ω) per square, is a measure of the resistance of a material to the flow of electric current between opposite sides of its surface.

A material's resistance will vary and change in accordance with the length, width, and thickness of the material. For example, a longer length of conductive thread will have more resistance than a shorter one, and, typically, thicker conductive threads have less resistance than thinner. Always check the material's specification sheet to determine the surface resistivity. If it's not available, you can use a multimeter.

So why not purchase the least resistive thread and fabric? Simply put: There's a trade-off between a material's physical properties and electrical properties. For example, stainless-steel conductive thread made entirely from 100-percent stainless-steel fibers can be more conductive than a similar-weight silver conductive thread made from nylon twisted with silver fibers. The advantage of stainless-steel thread is its high conductivity and tarnish-resistance. The disadvantage is that the stainless thread is less docile and more difficult to work with.

142 FOCUS ON TEXTILES

Textiles have been used by humans since prehistoric times—our first dyed flax fibers (dating back to 34,000 BCE) were discovered in a cave in the Republic of Georgia. Textiles are formed by weaving, knitting, crocheting, or pressing fibers together, and come from four main sources: plant, animal, mineral, and synthetic. Today, our textile choices range far and wide, from denim to leather to PVC and conductive fabric. This variety of makeups creates an infinite array of possible maker projects.

PVC FABRIC, A.K.A. VINYL, usually consists of a backing woven from polyester fibers with a surface coating of shiny plastic.

COTTON HAS BEEN INFUSED WITH NANOPARTICLES OF ZINC OXIDE TO CREATE A NEW TEXTILE WITH ANTIBACTERIAL PROPERTIES.

CHECK THE CONDUCTIVITY OF FABRIC USING A MULTIMETER.

Raid your closet for a old jeans, cut them into strips, and soak them in borax and water—you can use the resulting material as insulation. When the denim dries, it becomes resistant to fire, insects, and mildew.

TO SEW PVC WITH A SEWING MACHINE, USE A SHARP SIZE 11 NEEDLE. FOR HAND SEWING, USE A LEATHER NEEDLE.

VEGETABLE-TANNED LEATHER CAN BE OILED TO IMPROVE ITS WATER RESISTANCE.

BECAUSE PVC STRETCHES, THE THREAD YOU USE TO SEW IT ALSO MUST STRETCH. DON'T USE COTTON THREAD—USE THREAD THAT IS AT LEAST 60 PERCENT POLYESTER.

CARBONIZED SEWING THREAD WAS THE FILAMENT USED IN THOMAS EDISON'S ORIGINAL LIGHTBULB.

TO SOOTHE YOUR WEARY MAKER MUSCLES, CUT 1 FOOT (30 CM) FROM THE LEG OF AN OLD PAIR OF JEANS, SEW ONE END SHUT, FILL WITH RICE, AND SEW THE OTHER END. NOW YOU HAVE A MICROWAVEABLE HEATING PAD. JUST MAKE SURE THE DENIM YOU USE DOESN'T HAVE ANY METAL FEATURES.

REPURPOSE AN OLD PAIR OF JEANS INTO A BASIC ROLL-UP WRENCH SET (OR OTHER TOOL) HOLDER BY CUTTING OFF A LEG UP BY THE CROTCH SEAM AND CUTTING UP THE INSEAM TO MAKE A FLAT PIECE OF FABRIC. FOLD IT IN HALF, THEN LAY YOUR TOOLS ON IT, MAKING SURE TO LEAVE ROOM BETWEEN THEM (TO ENABLE ROLLING UP WHEN DONE), AND MARK OR PIN BETWEEN THEM. REMOVE THE TOOLS AND SEW A STRAIGHT LINE BETWEEN EACH TOOL POUCH.

TO PRACTICE WORKING WITH LEATHER, BUY SCRAP LEATHER BY THE BAG OR DECONSTRUCT ITEMS FROM A THRIFT STORE.

USE A WOODBURNING-TOOL TO ADD DECORATIVE DESIGNS TO LEATHER.

143 EXPERIMENT WITH FLEXIBLE SWITCHES AND SENSORS

Flexible switches and sensors can be crafted by combining conductive and nonconductive fabrics and threads. Hooks, snaps, beads, zippers, magnetic jewelry clasps—essentially any object that is metallic and conductive—can also be combined with conductive fabrics and threads to create switches or novel connectors. Below are some examples.

Ⓐ SOFT PUSH BUTTON Made from two pieces of conductive fabric separated by a perforated nonconductive layer, such as foam. The thickness of the nonconductive layer determines the amount of pressure required for activation.

Ⓑ FLEX SENSOR Constructed from pressure-sensitive resistive fabric layered between two pieces of conductive fabric or thread.

Ⓒ LINEAR SOFTPOT A potentiometer constructed from a layer of resistive fabric separated from conductive fabric with a perforated nonconductive spacer fabric layer.

Ⓓ CAPACITIVE TOUCH SENSORS With the help of a microcontroller, a piece of conductive fabric can be used as an electrode to sense touch or proximity.

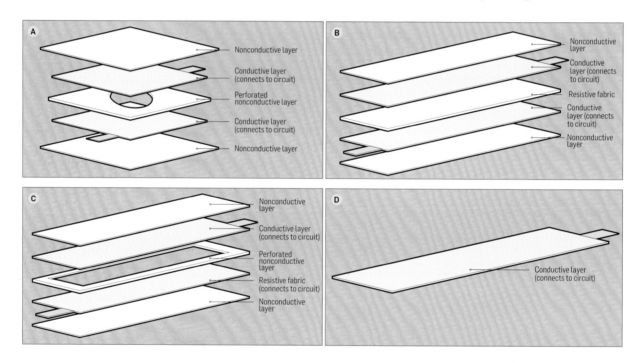

144 USE A MULTIMETER TO CHECK MATERIAL RESISTANCE

We've already discussed the importance of the material's resistance to the flow of electric current. Here's how to use your trusty multimeter to determine that value.

STEP 1 Locate the resistance function on your multimeter, typically marked by the symbol Ω. For ranging meters, set the range to 200 Ω. Lay a 1-foot (30-cm) length of conductive thread or a 1-by-1-inch (2.5-by-2.5-cm) square of conductive fabric on a nonconductive (nonmetallic) surface.

STEP 3 Hold each probe on opposite ends of the thread or square to get a reading. If your meter reading is zero or close to zero, select a lower resistance range. If the reading is 1 or displays "OL," select a higher resistance range until you get a readable value.

145 SEW A SIMPLE SOFT-CIRCUIT CUFF

The best way to see how soft circuits work is to just go ahead and sew one. Here's a basic circuit that only requires three LEDs (LilyPad sewable LEDs are a good bet), a sewable CR2032 coin-cell battery holder, a CR2032 coin-cell battery, conductive thread, a metallic snap, heavyweight nonfraying fabric such as felt, a sewing needle, and a glue gun.

STEP 1 Cut a 2-by-10-inch (5-by-25-cm) rectangle from the fabric. You can adjust the length so it fits around your wrist.

STEP 2 Following the circuit diagram, position the LEDs, battery holder, and snaps on top of the fabric. You can use a dab of hot glue to temporarily hold them in place. Make sure that you have positioned the LEDs in parallel. The positive leads of each LED need to connect and be sewn together. Likewise, all the negative leads need to be sewn together.

STEP 3 Cut a 2-foot (60-cm) length of conductive thread and thread your needle. Begin by sewing the positive terminal of the battery holder to the positive leads of the LEDs. Knot and cut your thread.

STEP 4 Sew the negative terminal of the battery to the male half of the snap. Knot and cut your thread.

STEP 5 Position the female half of the snap on the bottom of your fabric, then sew it to the negative leads of the LEDs. Knot and cut your thread.

STEP 6 Place the battery into the holder and snap the bracelet closed. Your LED lights should turn on.

MAKE YOUR OWN

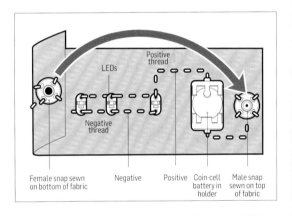

"Bright plumage displays are at the core of many mating rituals. Be the top peacock by wearing stuff that actually glows—this DIY light-up wrist cuff is your head start."

146 KNOW YOUR PHOTODETECTORS

Components that convert light into electrical signals come in a pretty mind-blowing variety. Bone up on the terminology and the standard package shapes, and you can at least avoid embarrassing yourself at cocktail parties.

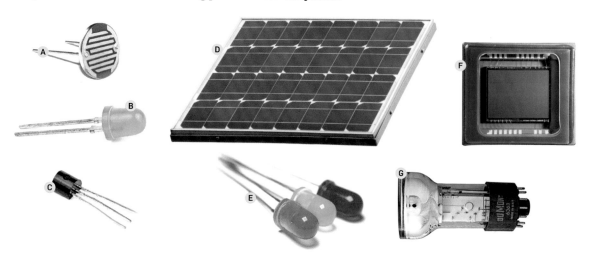

A PHOTORESISTORS Also known as photocells and light-dependent resistors, photoresistors have a high resistance in the dark that decreases as things get brighter. They're usually round with an interlocking "fingers" pattern on their surface, and they often have a red or pink color from the cadmium compounds they contain. Cadmium is really not good for you (or the planet), so resist the natural temptation to eat any photoresistors you may come across.

B PHOTODIODES Photodiodes can be wired in two modes: photovoltaic and photoconductive. In photovoltaic mode, the diode is connected in the so-called normal direction, with the anode positive and the cathode negative, and generates a current when exposed to light, like a solar cell. In photoconductive mode, the diode is connected backward or reverse-biased, with the cathode positive. In this mode, it blocks current, at least until it's exposed to light. A reverse-biased photodiode is like a light-sensitive valve for current from some other source.

C PHOTOTRANSISTORS They are easily confused with photodiodes, both in theory and in practice. Theoretically speaking, you can think of a phototransistor as a photodiode with a built-in amplifier; the practical upshot is that phototransistors respond to light more strongly but also more slowly. They can have two or three leads.

D SOLAR CELLS You've seen these giant photodiodes designed to operate in photovoltaic mode to generate power. But they can also be used as sensors, albeit very slow ones (at least on an electronic time scale). Like an alkaline cell, an individual solar cell has a characteristic voltage output and a set current capacity. They can be hooked together in series, parallel, or hybrid arrangements to make solar panels.

E LEDS Crank an electric motor with your hand, and you can use it as a generator. Shout into a speaker, and you can use it as a microphone. Turns out a lot of transducers (doodads that turn one kind of energy into another) can be used backward like this, and LEDs are no exception. They need to be reverse-biased in your circuit and will need some kind of amplification, but there are advantages: LEDs are dirt cheap, available pretty much everywhere, and selectively sensitive to the wavelengths of light they're designed to emit, so you can use them to detect colors.

F IMAGE SENSORS These are integrated circuits with grids of very many, very small photodetectors baked right in. For visible light, there are two general types—the charge-coupled device (CCD) and the active pixel sensor. The difference between them has less to do with the pixels themselves and more to do with how image data is extracted from the grid once captured. CCDs are a more expensive, older technology and are still favored for high-quality imaging applications. But the gap is closing.

G PHOTOMULTIPLIERS Based on vacuum-tube technology, photomultipliers today mostly find use in high-end defense, industrial, and research applications. They are expensive but offer an unparalleled combination of sensitivity and speed. Some can even detect single photons.

147 IMPROVISE A PRESSURE SENSOR

Not to be confused with a pressure switch, which is just on or off, this analog device has a resistance that changes smoothly with the amount of pressure applied. This roll-your-own version is a classic hack from Forrest Mims. (You do know who Forrest Mims is, don't you? If not, go find out. Right now.)

STEP 1 Snag a piece of the antistatic foam that chips (the inedible kind) are often packaged in. You can cut it to pretty much any size or shape you need.

STEP 2 Strip the ends of two wires and poke them all the way through the foam on opposite sides. Bend the protruding ends over so they don't pull out.

STEP 3 Dip or paint the whole shebang with liquid electrical tape or PlastiDip, let it dry, and your sensor is ready to use. Connect to a microcontroller and calibrate.

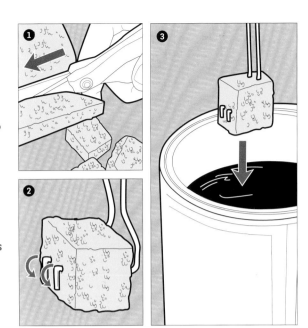

> "Sensors try to seamlessly integrate electronics into your life. Think of nontraditional switches as a way to further your transition to robot."

148 RIG A LOOP-SWITCH BOOBY TRAP

The loop switch may be the world's simplest sensor. It takes 15 seconds to slap one together from two pieces of insulated wire, and it's reliably closed by pulling it taut. Use it to activate a burglar alarm to keep people out of your secret stuff, to set off a camera, or to find out who's come a-snoopin'.

STEP 1 Strip 1 to 2 inches (2.5–5 cm) of insulation off one end of a piece of wire. Form the stripped part into a loop big enough to easily pass over another piece of the wire, and twist to secure.

STEP 2 Pass a second piece of wire through the first, strip the end again, and form another loop, this time being sure to include the first wire inside the second loop. Twist to secure, leaving plenty of room for the wires to slide back and forth.

STEP 3 Electrically connect the two wires between a power source and the load you want to activate, and mechanically between a stationary object and the door, drawer, window frame, or other movable object you want to secure. Then wait for someone to trip it!

149 DECIPHER DECIBELS

Basically, 50 to 60 decibels is a normal speaking voice. Power tools are around 90 to 100 decibels. Drugstore personal alarms advertise 120 decibels, and that's louder than anything you'll need to build. Tiny smoke alarm piezos reach 90 decibels—with those, you should have no trouble getting attention.

150 GET VARIOUS NOISES FROM A BUZZER

If you're looking for the dirt-simplest way to make sound happen, you'll want to look for a piezo buzzer with a built-in driver. These aren't hard to find and may be labeled "pulsing buzzer" or similar. A built-in driver provides a fixed-frequency signal, so all you have to do is plug in power.

So let's make some noise with it. All you'll need is a buzzer with a built-in driver (rated for at least 10 volts), a few electrolytic capacitors (22 µf to at least 220 µf), a 9-volt battery connector with leads, a piece of breadboard, a 10k-ohm potentiometer, and a light-dependent resistor (a.k.a. LDR or photocell).

A Hook the buzzer up to the battery. You will hear the irritating whine of success. Plain and simple.

B Time to make it less lame. Plug in a 220-µf capacitor in parallel with the buzzer. See what happens when you remove the battery? Test it with another value. For a good practical lesson in buzzers and capacitors, find a 2,200-µf capacitor and try it out. You'll notice the volume fading and the pitch rising slightly as the capacitor drains. As the voltage varies, so does the oscillation frequency.

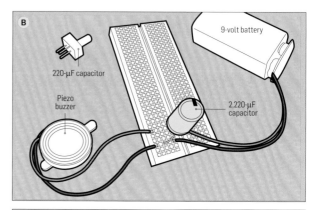

C To add the volume control you wish you had in the first build, add a 10k-ohm potentiometer to the circuit. The battery should connect to each end of the pot. Connect the buzzer's positive lead to the pot's middle pin, and connect all the negative ends together. Now turn it down a little. Much less likely to infuriate the NIMBYists.

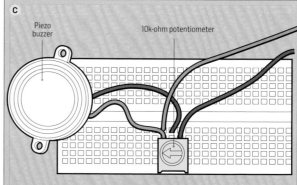

D This one's no more complicated than the rest but much more entertaining. Hook the negative ends from the buzzer and battery together. This time, instead of connecting the positive wires directly, put an LDR in between them. Try it out. You can control the sound with your shadow. Not too shabby for just three components.

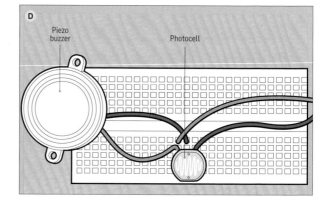

151 WORK WITH SPEAKERS

Adding lights to your project is good solid making, but adding sound can result in quite a big bang. Here's a general overview of four main types of speakers to choose from, depending on what you're after.

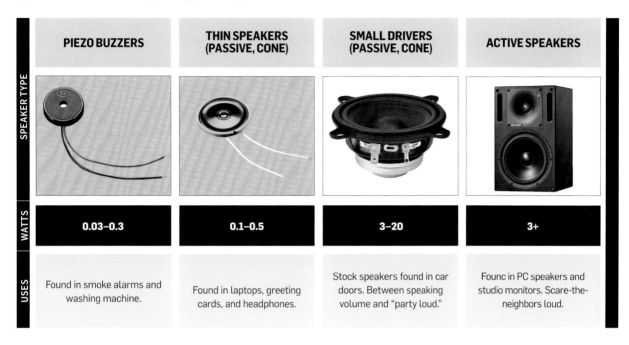

SPEAKER TYPE	PIEZO BUZZERS	THIN SPEAKERS (PASSIVE, CONE)	SMALL DRIVERS (PASSIVE, CONE)	ACTIVE SPEAKERS
WATTS	0.03–0.3	0.1–0.5	3–20	3+
USES	Found in smoke alarms and washing machine.	Found in laptops, greeting cards, and headphones.	Stock speakers found in car doors. Between speaking volume and "party loud."	Founc in PC speakers and studio monitors. Scare-the-neighbors loud.

152 BUILD A BASIC DIY AMP

Let's face it, there's no point in building a noisemaker circuit if no one can hear it. Fortunately, the LM386 is a great little chip to employ. It's perfect for portable, battery-powered amps, and has enough power to bump up to turn-it-down loud.

Essentially, a very small signal is used to control the flow of the power supply through to the output. Some LM386s can run off as much as 18 volts, but 9 volts is a pretty powerful, safe amount, and the batteries are easy to find.

This is a class-AB power amp, which basically means it boosts up the volume with a beautifully small number of components, but with a perfectly acceptable level of loss and distortion. Also, it's a single-channel amp, so if you want stereo sound, you'll need to double it up.

Assembled as shown in the circuit diagram, the amp will provide a gain of 20, which is pretty good. However, adding a 10-uf capacitor between pins 1 and 8 will boost it all the way up to 200. Add a resistor in series with the new capacitor to tone it down. The beauty of audio circuits is that they're so easy to play with. Try swapping in components of different values until it sounds right to you.

Pro tip: Datasheets are an invaluable source of reference schematics and advice.

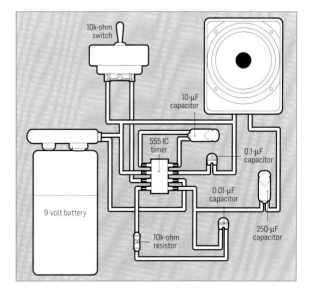

153 MASTER MICROPHONES

On a fundamental level, microphones are just speakers working in reverse. Practically speaking, though, you'd be hard pressed to find a microphone turned into a decent speaker. Here are a few types of common mics to consider. Electret microphones are a type of condenser mic, and if you're shopping around for bare components, it's the most common type you'll see.

THINGS TO LOOK FOR

DIRECTIONALITY Omnidirectional versus unidirectional
FREQUENCY RESPONSE Are you capturing voices, thunder, or a bit of everything?
SENSITIVITY Usually larger = more sensitive. Keep in mind that a preamp can do wonders for a weak signal.

MICROPHONE TYPE	PIEZO	ELECTRET	DYNAMIC
BENEFITS	Needs no power; is thin; can double as a knock sensor.	Needs power; has good frequency response; may require a preamp; is sensitive; can be quite small.	Needs no power; is rugged; usually works without preamp.
USES	Found in percussion and experiments.	Found in phones, laptops, and headsets.	Found in studio recording and stage audio.

154 MAKE A PIEZO MIC

Want to experiment recording percussion on instruments and random objects? Try a piezo contact mic.

STEP 1 Take a mono audio cable. Cut and strip one of the ends.

STEP 2 Obtain a piezo. You can extract one from an old noise-making toy or buzzer.

STEP 3 Solder leads to the corresponding negative and positive parts of the cable. **Pro tip:** A naked piezo may be challenging for beginners to solder to. Just give that wide metal area plenty of time to heat up.

STEP 4 The leads will need a strong mechanical connection to hold them in place. A dab of hot glue works for everything. Then seal everything else with tape, glue, heatshrink, or something you scrounge up.

STEP 5 Use a rubber band or clamp to attach the piezo's flat side to a noisemaker—such as an improvised drum made out of a can—and go nuts with the new sounds you make.

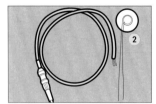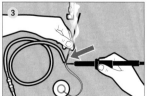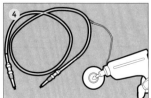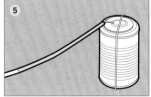

155 BEND THAT CIRCUIT

The antithesis of a conventional step-by-step project is the art of circuit bending, dismantling low-voltage, battery-powered audio toys and altering their circuits to create new, unexpected sounds. Notoriously, the art form was born in 1967, in the workshop of Reed Ghazala, when a toy circuit in his desk drawer shorted out and started repeatedly making all manner of strange oscillating sounds. Ghazala wondered, "If that can happen by accident, what can happen on purpose?" The name alludes to the mind-bending experience that can result.

Pivotal to circuit bending is embracing experimentation. There's no prescription, no set way of doing things, only play. A trip to the thrift store can yield a wide array of musical toys on the cheap, ready to be taken apart and altered. Pair these finds with a selection of resistors, potentiometers, switches, knobs, and buttons, and the result is anyone's guess.

For starters, though, you can play open circuits. Basically, open up your toy, reveal the circuit, and use alligator clips or your hands to play with the board and see what sounds it can make. Once you find points that make sounds you like, you can wire the points to a switch, so you can hit those sounds on demand.

Word of advice: Don't attempt to circuit bend anything that plugs into a wall. High voltage is not what you want to have for breakfast.

156 GET TO KNOW RADIO

Electrons flowing in only one direction through a circuit are called a direct current (DC), while electrons wiggling back and forth are called an alternating current (AC). Unlike continuous DC, AC has a characteristic frequency describing how fast the electrons change direction. At frequencies above 3,000 wiggles per second or so, circuits carrying alternating currents emit a kind of electromagnetic (EM) radiation that we call radio waves. A circuit designed to do this on purpose is called a radio transmitter.

Like visible light, infrared light, X-rays, and other kinds of EM radiation, radio waves have measurable properties including wavelength (the distance between the highest points on adjacent waves) and amplitude (how strong or tall the waves are). When a radio wave strikes an object made from metal or another conductor, the electrons in that object wiggle back and forth sympathetically, producing an alternating current that can be detected by a connected circuit called a radio receiver. Any variations in the frequency or amplitude of the current that produced the transmitting wave will be reproduced in the current picked up by the receiver, and these variations can be used to encode (or modulate) information.

That these things happen is, of course, extremely useful to smart monkeys like us. The oldest and most fundamental application of radio technology is to instantly (OK, OK, almost instantly) send information over large distances—whether around the block, across the ocean, or out into interplanetary space. Newer, but perhaps equally important, is the use of radio waves as radar (originally an acronym for "radio detection and ranging," coined by the U.S. Navy in 1940), to locate distant objects like airplanes and thunderclouds. Nowadays, we often use radio over very short distances for simple convenience, to save time and trouble that would otherwise be required to connect nearby devices with a wire.

157 LISTEN IN WITH A TRENCH RADIO

Believe it or not, radio wasn't born in the dashboard of your dad's car—and it didn't die there, either. Soldiers used to set up these simple receivers in the trenches to follow the news, and it'll still do the trick today.

STEP 1 Use a safety pin to poke a hole in a toilet-paper tube, and secure a magnet wire to the tube by tying one end through the hole.

STEP 2 Create a coil by wrapping the magnet wire tightly around the tube 120 times, making sure that the wire is packed closely together as it coils. The number of coils affects what radio stations you'll pick up, so experiment with their arrangement if you aren't hearing your desired station clearly.

STEP 3 Make a "cat whisker" by poking the safety pin into the graphite of a pencil stub's dull end.

STEP 4 Hang an antenna cable far out the window and attach a ground cable to a metal radiator inside your home. If you don't have a radiator, attach the ground cable to a metal coat hanger and stick it in the ground outside.

STEP 5 Place the toilet-paper tube and a blued razor blade onto a wood board and push in two thumbtacks next to the tube and two next to the razor blade. Wrap the antenna wire around the components according to the diagram, twisting wires together.

STEP 6 Peel back the insulation on your headphones' audio jack. Use some stripped wire to connect the safety pin in the pencil's lead to one of the audio jack's wires. Then connect the audio jack's other wire to the ground.

STEP 7 Don your headphones, and touch the pencil lead to the razor blade. Move the pencil around until you pick up some good news.

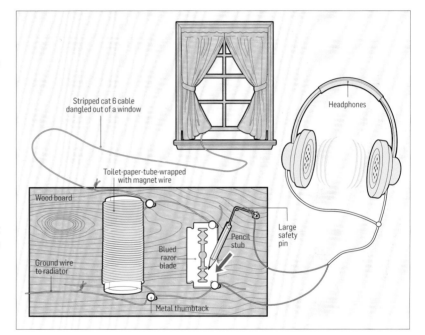

158 RIG A CELL-PHONE BLOCKER

Privacy has become a hot commodity nowadays. To fly under the radar and prevent people from picking up your mobile GPS coordinates, tightly wrap your cell phone in multiple layers of aluminum foil. This is a type of Faraday cage—an enclosure that blocks electrical fields by distributing incoming and outgoing charges within the conducting material.

Traditionally used to protect electrical equipment from lightning strikes, a Faraday cage can also block the radio waves that cell phones need to communicate, which includes Internet connections and GPS tracking signals. To test your Faraday cage, try calling your phone; if your call goes straight to voicemail, it means that your enclosure works and signals to and from your phone have been successfully blocked.

159 GIVE A HOMOPOLAR MOTOR A SPIN

The so-called homopolar roller, though not the simplest motor to understand, is hands down the easiest to make. You just need four parts: a 5-inch (12.75-cm) piece of copper wire, a AA battery, and two round rare-earth (a.k.a. neodymium) magnets.

STEP 1 Bend the wire into a U shape and strip both ends.

STEP 2 Snap a supermagnet onto the battery at each end and lay it down on a flat surface.

STEP 3 Bend the stripped ends of the wire a bit to match the radius of the magnets. You want a constant sliding contact between the wire and the magnets as they turn.

STEP 4 Connect the wire across the magnets and let go!

MAKE YOUR OWN

160 WIRE UP A REVERSE SWITCH

Say you're building a widget with a DC motor—a toy, a robot, a tool, whatever—and you want it to be able to run forward as well as backward. So you've got a DC motor and a battery with leads. Wire the leads to the motor terminals, and it spins one way. Swap them, and it spins the other. But you can't stand there swapping wires by hand all day. Here's how to do it with a switch.

STEP 1 Grab a double-pole double-throw (DPDT) switch of the "center-off" variety. It has six contacts: two common "input" terminals, where you attach your power leads, and four switched terminals—one pair for each "output." The common terminals are almost always the two in the middle. "Center-off" just means that the switch has a third, middle position where none of the leads are connected—handy so you don't have to wire a separate power switch.

STEP 2 Solder the leads from your power source to the two common terminals. It doesn't matter which one goes to which side.

STEP 3 Connect the leads from your motor to the switched terminals on one side of the common contacts—it doesn't matter which. You're wiring the switch, arbitrarily, so the motor spins one way.

STEP 4 Cut, strip, and connect two short crossover jumpers between the pair of unused contacts and those that connect to the motor. The jumpers should be in an X configuration, with each one crossing over to the opposite side of the switch. There you go—toggle the switch to make your motor spin in one of two directions.

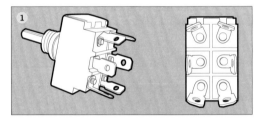
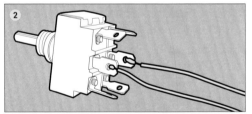
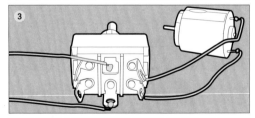
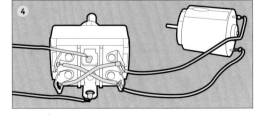

161 MACGYVER A GENERATOR

Whether you're camping, moving off-grid, or surviving a minor (or major) apocalypse, you may be called upon to run a refrigerator, lights, or other AC appliances away from a source of mains electricity. AC motors in general can be run backward to generate power, and AC induction motors are especially good for this. (By "backward," I mean turning the shaft to generate electrical power instead of applying electrical power to turn the shaft.) If you do this, you should turn the shaft in the same direction it turns when running as a motor. And always respect the power you generate as much as you would mains electricity—this stuff can kill you as sure as what comes out of your wall outlets.

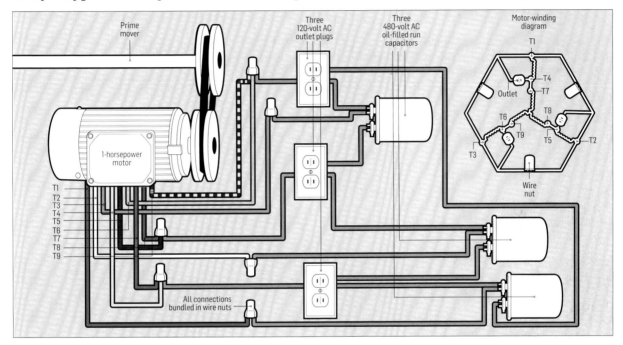

STEP 1 Beg, buy, scrounge, or steal (OK, don't steal) a 1-horsepower motor. The plate on the housing should identify it as three phase—208- to 230/460-volt AC, 1,700 to 1,800 RPM—with nine leads, wired in a Y or star configuration. Run as a generator, this machine should be good for about 500 watts.

STEP 2 Open the wiring box on the motor's side and connect three capacitors and three 120-volt AC outlet plugs, as shown in the circuit diagram. You can use polarized or grounded receptacles, but be aware that there's no neutral wire or ground connection in this arrangement, as in household wiring. So don't try to connect it to any standard electrical grid. Use 12-AWG insulated stranded wire with wire nuts and crimp-on terminals for the caps. Make sure everything is safely insulated.

STEP 3 Hook up a gas engine, wind turbine, or other prime mover by belt drive. Induction motors are asynchronous, meaning the shaft must turn at a slightly different speed than the rotating magnetic field that drives it. This difference is called slip, and it's typically 3 to 5 percent. The practical upshot is that an induction motor that is marked 1,800 RPM actually turns at about 1,725 RPM and, when used as a generator, needs to be driven at about 1,875 RPM. Start the shaft turning, then briefly plug in an incandescent lightbulb to verify that you're getting power.

STEP 4 If you've got it hooked up right and aren't getting joy, your motor may need to be briefly flashed with DC current. Induction motors don't have permanent magnets and depend on external voltage when starting to generate a magnetic field. Frequently used models usually become magnetized and will start generating without help; those that are new or have sat for a long time may need to be flashed. To do so, start it turning, then briefly connect a car battery across one of the outlets for no more than a second.

STEP 5 Once the generator is turning and making watts, plug in whatever it is you want to power, then use your multimeter to check the voltage at the outlets. Adjust your drive speed to get it as close to 120 volts AC as you can.

162 STRIP A DRILL FOR PARTS

Cordless drills contain high-power DC motors with compact planetary gear transmissions—and they make perfect robot drive motors. If you have an obsolete or broken drill, it's very much worthwhile to strip it for parts.

STEP 1 Remove the battery. These drills have quite a kick, both mechanically and electrically. Safety (more or less) first.

STEP 2 The chuck is likely secured with a left-hand threaded screw or bolt through the spindle's bottom. Open up the chuck jaws and look inside with a flashlight. Loosen the screw head, tap the bolt out, and then remove the chuck assembly by turning it counterclockwise off the spindle. **Pro tip:** A heavy (³⁄₈-inch [8-mm] or ½-inch [9.5-mm]) Allen wrench can be clamped in the jaws and used as a lever to loosen a stubborn chuck.

STEP 3 Remove any screws and open up the case.

STEP 4 Cut or desolder the motor wires. The trigger switch and reversing switch may also be worth salvaging.

STEP 5 Remove the clutch. If it's a high-end drill, the drivetrain may be nicely modular, in which case you may only have to lift it out, leaving the motor and planetary gearbox for you to feast upon. If it's a knock-off, the clutch and gearbox may be a single assembly, in which case you'll likely have to remove some screws and pop out some tiny fussy parts, like individual ball bearings.

STEP 6 A planetary gearbox is driven by a small central gear (a.k.a. the sun gear) connected to the motor shaft. That's your "input" torque, which is transferred by the orbiting planet gears to the outermost annulus gear, which is a ring with teeth on the inside. Depending on your make and model, removing the clutch may have left the annulus free to rotate inside its housing without any convenient means to attach a wheel or other "output" load, so the final problem is to fix things so that the annulus and spindle rotate as one. Opening up the gearbox and squeezing an O-ring in on top can work, and it provides handy slip-before-it-strips functionality. Or if you need the full torque, guaranteed, you can drill and tap for short screws.

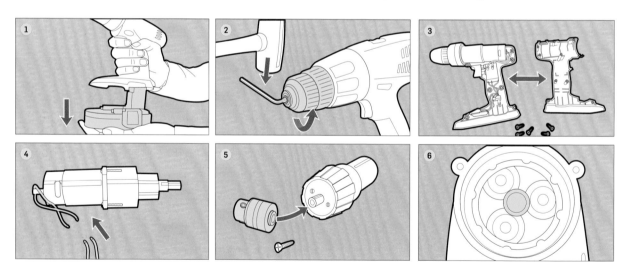

"For most people, a cordless tool is trash once its battery refuses to take a charge. But for the real maker, the party's just getting started."

163 ROLL YOUR OWN GEARS

Gearmaking is an exacting science. For heavy machinery or other high-torque loads, it makes more sense to use manufactured gears, available in every configuration.

Still, there's something hugely satisfying about seeing a homemade gearbox come to life, and if you're working with light torques in an easy material like wood or plastic, it's not difficult to do it yourself using one of several methods.

Your first problem will be designing the profile of the gear, but modern software makes this easy. The free open-source drawing program Inkscape is a good place to start. It even comes with a rendering extension that generates vector art for simple gears when you input the diameter, the number of teeth, and the tooth angle you want.

HAND CUTTING With patience and decent tools, gears can be cut by hand. You need an accurate pattern to follow, but pretty much any modern printer is capable of $1/1000$-inch (0.025-mm) accuracy or better. Print your pattern on a mailing label, fix it to your stock, and cut to the lines with a scroll saw or fretsaw, followed by light filing.

CNC ROUTING A router is a lot easier to come by than a lasercutter, but the "pointiness" of the inside angles you can cut is limited by the radius of the rotating physical tool itself. You won't be able to cut teeth narrower than the diameter of your cutter, often ¼ inch (6.5 mm). (See #173–175 for more information.)

LASERCUTTING You likely don't own a lasercutter, but a hackerspace near you may. Plus, there are tons of folks competing for your lasercutting dollars online. Just upload your file and your parts arrive a few days later. **Pro tip:** Ensure your dimensions and units get preserved—or you can have, say, a part designed in centimeters come back cut in inches. (See #188–191 for more lasercutting info.)

3D PRINTING Even low-priced printers can make plastic gears accurate enough to be used in light loads. Unlike other methods, 3D printing breaks you out of flatland so you can create bevel, helical, or worm gears. You'll need a full 3D model. (See #176–178 for 3D printing info.)

164 SHIM A GEARBOX

Whether it's in a car or a toy, almost any off-the-shelf gear train can be improved by a good shim job. If you're a performance nut, this may be worth doing for its own sake. Even if you're not a performance nut, all transmissions wear with use, affecting how well the teeth mesh and how much play there is in the mechanism. If gears mesh too tightly, their action will generate a lot of heat and wear quickly; if it's too loose, the action will be loud, inefficient, and prone to slip.

The shims themselves are just very thin pieces of sheet metal cut (natch) from "shim stock," which ranges in thickness anywhere from $1/1000$ to ¼ inch (0.025–6.5 mm). It can be a pain to cut shim stock without raising a burr that screws up its nominal thickness, so you're usually better off buying shims than trying to make your own.

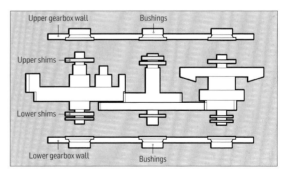

To use, disassemble your gear train and add shims over the shafts between the gearbox walls and the gears, as needed, to nudge them over where you want them to be and keep them there. Start shimming at the input gear and work your way through the transmission to the output gear. Otherwise it's pretty much a trial-and-error process that can get pretty tedious depending on how many gears are involved.

165 PICK A POWER SOURCE

Whether you're trying to power the hideout that will help you stand a fighting chance of surviving the zombie apocalypse, or just trying to test out the circuitry for your newly wired automated cat feeder, having a variety of power sources at your maker command can only be a good thing.

SOLAR PANELS Our way of harnessing the sun's energy and transforming it into electrical energy. No doomsday bunker or off-grid hideaway is complete without these.

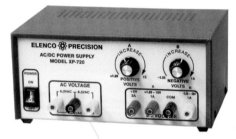

BENCH POWER SUPPLY Useful part of a maker's arsenal for its ability to adjust output voltage over a certain range. Comes in handy when trying to test circuitry without frying it. Relatively inexpensive, or free if you build one from obtainium parts.

COIN-CELL BATTERIES Essentially the same in function as household batteries. Most often used in devices that don't require much power but do require power for an extended period of time. They'll power your watch, calculator, and eventually your hearing aids—a lot of times for a year or more.

AA BATTERIES Like mini portable power plants, turning stored chemical energy into electrical energy. Popping a battery into your TV remote closes an open circuit, changing your channels as restless electrons rush from the anode to the cathode.

PERSON-POWERED GENERATOR Basically a motor working in reverse—current is created when some force causes conductive wire to pass through a magnetic field. Powered by the initial energy of however long you can pedal or crank.

PELTIER JUNCTION Thermoelectric modules that, when connected to electricity, function as a heat pump, heating one side and cooling the other. Can also generate energy manually—the transfer of heat between the two sides creating current in the process.

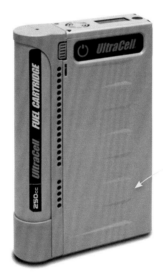

FUEL CELLS Like batteries in the sense that they transform chemical energy into electrical energy. The main difference is that fuel cells use fuel stored in an external source. Where batteries run dead when internal stores deplete, fuel cells run as long as there is fuel.

BATTERY CHARGERS A lot of batteries are rechargeable, but before you start recharging anything, make sure your charger and batteries are made for each other. Not all batteries (or chargers) are created equal, and nobody but the Internet likes an unintended car-battery explosion.

PHASE CONVERTER Converts three-phase electric power to single-phase electric power, or vice versa. Three-phase power can be used to power heavy loads (like electric trains) or transmit power.

HACKETT SAYS

166 HARVEST ELECTRONIC OBTAINIUM

Our consumer economy is rooted in built-in obsolescence. Your shiny state-of-the-art device is manufactured to look dated and cheap in a few years, and it will probably fail altogether in three. As a maker, there is only so much you can do—repair all you want, but between sealed, black-box construction and the death of support media (only us olds even know what a cassette tape is; soon CDs and DVDs will join its ranks), your fancy bit of high tech will be still-shiny trash.

Recycling electronics is better than trashing them, but even better is treating them as a source of obtainium—a font of future parts for future projects. There's low-hanging fruit that's easy to pluck and well worth the effort: plugs with their cords still attached, speakers that still make sounds, and low-voltage DC computer fans that just slide right out of their former homes. But also keep an eye out for the good stuff on the higher end, which tends to be harder to remove though worth it. For instance, the rare-earth magnets deep inside hard drives and the heatsinks stuck to circuitry are strangely expensive, so dig deep to score them. Scavenged LCD panels and LED arrays will look downright fancy on your own builds—just be sure to keep the ribbon and wiring intact. Pull apart larger copiers and printers to mine old gearing, shafts, and linear actuators; plus, the stepper motors lurking in these machines are great for small generators, robotics, and DIY CNC machines. Here, a motor salvaged from an old scooter is on its way to becoming part of a hacked-together bicycle-powered generator. Obtainium—the gift that keeps on giving.

167 COOK WITH THE SUN

Solar power means more than just electricity from sunlight. Long before there were solar cells and panels, optical solar concentrators using nothing but mirrors and lenses were starting fires, melting metals, and driving engines. Today, there are solar furnaces that can produce temperatures approaching the surface of the Sun. Making meals is kid stuff by comparison.

STEP 1 Take a foil-lined car sunshade and attach a strip of Velcro along the edge to one side of the notch where the rearview mirror is supposed to go. Stick the mating Velcro strip to the opposite edge.

STEP 2 Bring the two edges together, reflective side in, to form a funnel. Secure with Velcro.

STEP 3 Set up the reflector on a bucket in a nice sunny spot, on a calm morning, and add a wire cake or grill rack to hold it down and provide support for the pot.

STEP 4 Mix your ingredients in a black pot, cover with a matching lid, put the pot in a plastic oven bag, and set the bag on the rack.

STEP 5 Wait. Solar cooking is slow cooking. Depending on what's in the pot and the weather, your dish may be ready in 2 to 8 hours.

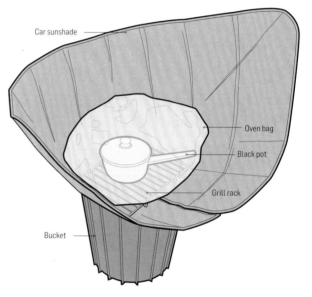

168 GET SOLAR PANEL SAVVY

Pick up a standard AA battery and look at it. Guess what? You're not actually looking at a battery, you're looking at a cell—specifically, an electrochemical cell. Connect many together to power, say, a flashlight, and the collection of cells altogether is, in truth, a battery.

Why all this talk about batteries in a discussion of solar power? Simple: If you understand the relationship between electrochemical cells and batteries, you understand the relationship between solar cells and solar panels. Both solar and chemical cells produce DC voltage across a pair of metal electrodes through a physical process: In a AA cell, it's a chemical reaction; in a solar cell, it's a photovoltaic one.

Just as there are several different chemicals that can make up a AA cell (alkaline, nickel-cadmium, etc.), there are several different semiconductors that can make up a solar cell (silicon, cadmium telluride, etc.). And just as a single electrochemical cell typically generates about 1.5 volts, regardless of size, so, too, does a single solar cell generate about 0.5 volts.

You can make these cells bigger or smaller, but, again, as with chemical cells, that only changes the amount of current that they can produce. Connecting cells in parallel—with one wire running between all their positive ends and another wire running between all their negative ends—is not very much different, electrically speaking, than creating a single cell that is physically larger.

To increase voltage, on the other hand, you have to connect cells in series, with the positive end of one connected to the negative end of another, and on and on as needed, with your circuit connected to the positive and negative ends of the whole stack. In this case, the voltages of all the individual cells add together. Assuming your solar cells put out about 0.5 volts, connecting 10 of them in series would give you 5 volts across the stack. If one stack can't produce enough current for your needs, well, you just make it bigger by connecting it, in parallel, to one or more identical stacks.

169 SAVE DAYLIGHT IN A JAR

There was a time in the not-too-distant past when the idea of a solar-powered flashlight was literally a bad joke. Not so much anymore. Advances in solid-state lighting have made it not only possible but commonplace for small, inexpensive electronics to store up energy from daylight, then release it as useful illumination when darkness falls. This DIY version is called a sun jar.

STEP 1 Cut a circle of perfboard the size of a large canning jar lid.

STEP 2 Mount two AA battery holders, a diode, a PNP transistor, two resistors, and a yellow LED on the bottom of the perfboard, and interconnect with jumper wires per the circuit diagram here.

STEP 3 Solder leads to a small solar panel and wire them to the circuit from the top side.

STEP 4 Insert two rechargeable nickel metal hydride batteries, test to make sure everything works, and then glue the solar panel to the top of the perfboard.

STEP 5 Put the perfboard on top of a canning jar and secure with the band. Set it up in a bright window, wait for night to fall, and enjoy. **Pro tip:** Etch or clear-coat the inside of the jar for a more diffused glow. For extra romance, use a flickering LED.

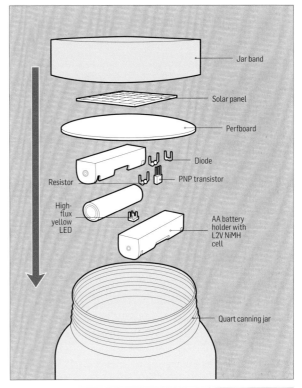

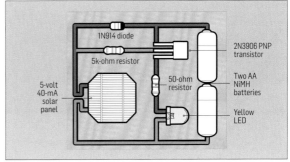

"All energy sources, including gasoline and fire wood, are really just pent-up solar energy. When you use a solar panel, you're cutting out millions of years of waiting for dead dinosaurs and plants to turn into fossil fuels."

170 SCHEME UP A WIND TURBINE

MAKE YOUR OWN

Wind is the perfect source of electricity for the DIYer, especially as part of a hybrid system. Not only can you get the obvious benefits of power at night or when it's cloudy, it's also completely customizable, depending on your demands and the amount of wind your location receives.

If all you need is something to charge a couple of batteries for lights and a little computer usage, you could make a wind generator almost completely out of scrap. If you have constant but low wind speeds, you can design something to start generating at just a couple of miles per hour. Alternatively, if your wind is superstrong and unpredictable, you need to make something to withstand the extra force. In order to design something that is exactly suited for you and your area, there are a few things you'll need to know.

TURBINE TYPES There are two main types of wind generators: horizontal-axis wind turbines (HAWT) and vertical-axis wind turbines (VAWT).

The most common is the HAWT. Most of the commercial home units you can buy are HAWTs, as well as those in the large wind farms and in water-pumping mills. The blades are perpendicular to the wind's direction, and a tail keeps the blades lined up. The blades are designed as an airfoil (like an airplane wing) to actually use lift, rather than drag, to make the blades move. This approach is very attractive, as your blades can travel faster than the wind speed.

VAWTs tend to be much sturdier than HAWT designs. Most work on the concept of drag instead of lift, so the blades don't need to be nearly as well made. However, the disadvantage is that speed and efficiency are often limited. This design is usually best suited to when your wind is strong with powerful gusts.

No matter what your design, the basic premise remains the same. Blades transform wind into mechanical energy, and this mechanical energy turns a generator that transforms the mechanical energy into electricity.

BLADES Blades have been made out of every material imaginable: metal, wood, plastic, fiberglass, even PVC.

With a VAWT, you basically just need to make something that will catch the wind—even barrels cut in half will do the trick. For more efficiency, use three or more blades with a slim airfoil.

In a HAWT design, your blades will want to be tapered, slightly twisted, and shaped into an airfoil to help with efficiency and speed. Wood and PVC are commonly used because they're easy to carve.

HUB/BLADE MOUNT Attaching the blades to your motor needs to be done through some kind of hub or mount. This piece receives a lot of force and needs to be strong, yet easy to dismantle for tweaks and tests. If your motor has a threaded shaft, you can thread the hub to the motor, but make sure that the blades do not spin in the same direction that the hub unscrews.

MOTOR There are many options for the motor. You can make one yourself, using magnets and copper coils. You can fine-tune the type of power produced, depending on the strength of the magnets, the amount of coils you use, and the configuration of the two together.

Alternatively, you can find a ready-made permanent magnet generator, like a treadmill motor. The trick with sourcing a motor is to find one that produces enough voltage to charge your batteries at low speeds. You should look for 1 volt per 30 RPM or less.

TAIL A tail can be simple. All it really needs to be is a piece of stiff material, something to keep the machine lined up in the wind. Tails should be made to hinge, so that they can turn the generator out of strong wind. You want the hinge off center and the tail weighted enough so that it can withstand normal winds. However, higher winds should be able to fold it back on itself.

DIODES AND WIRE You'll need to have the motor connect to a one-way diode. This will stop the motor from using the power in your batteries to make it turn when there's no wind. If your motor outputs AC power, you can use a rectifier as your diode.

The size and type of your wire depends on your setup and system voltage. There's a drop in voltage the further the electricity has to travel. To prevent this, you'll need to increase the thickness of your wire the farther away your wind turbine is from your batteries.

TOWER The simplest tower is a pipe with a hinge on the ground and four guy cables. This allows you to raise and lower the turbine with relative ease. The cables should be properly anchored with turnbuckles for adjusting tension.

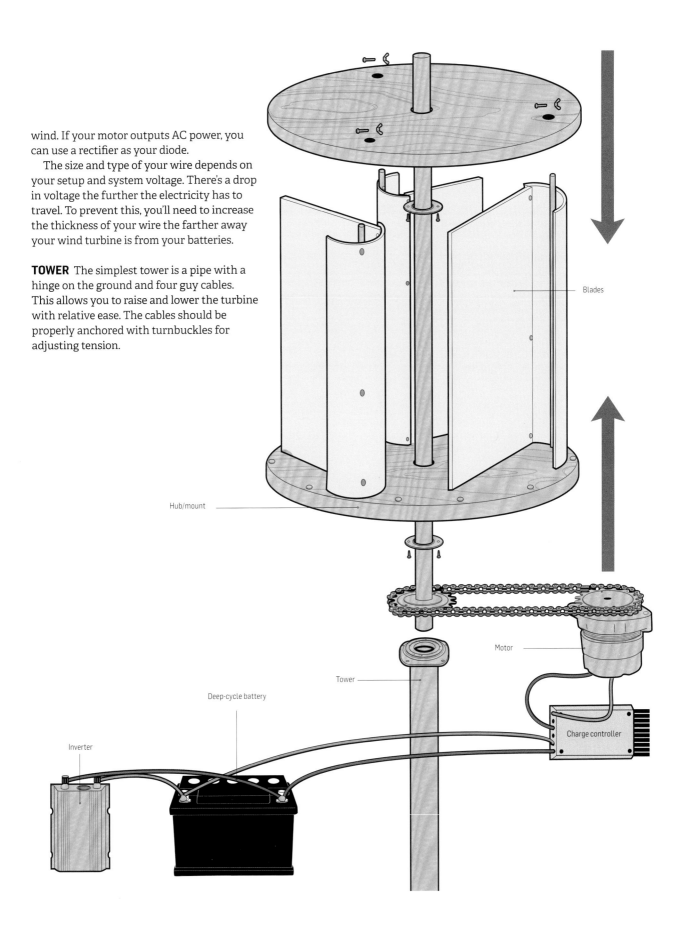

171 POWER UP WITH PEDAL POWER

At a steady pace, your average bicyclist could pedal out between 120 and 300 watts of power in a 20-minute session on this generator. Your typical 12-volt car battery holds up to 500 watts, so you could fully charge it (and get some serious cardio, too) with a little time on the bike. Most small household appliances don't draw a lot of power. Your laptop runs on 50 watts, so 20 minutes of cycling will run it for almost an hour. Lightbulbs start at 25 watts, but LEDs and fluorescents have a base of 5 watts—and give 2½ hours of light. Your coffeemaker needs about 130 watts to brew a full pot in 10 minutes. Consider your various power needs, and make sure your battery can handle several recharges.

"Someday soon, I will be a power baron: Drudges will sweat bullets to charge their phones on my bike, leaving a surplus for me to squander on high-res cat videos and ice-cold beverages."

172 BUILD A BICYCLE GENERATOR

Copper wire ties modern civilization together, but it is a fragile thread. In 2012, Hurricane Sandy bathed miles of the electrical conductor in saltwater, severing my Brooklyn neighborhood from the grid.

To prepare for the next disaster, I decided to build a simple, robust generator able to power essentials like a refrigerator, chargers, and lights indefinitely. Gas and diesel generators are common backup sources of electricity, but disasters can easily disrupt fuel distribution networks, as Sandy did. I decided that the best power source would be me: I'm always around, and my fuel is readily available. I just needed a means to turn calories into kilowatts.

Bikes efficiently convert muscle energy into rotary motion, which can rapidly spin a generator. I extracted a motor from a trashed mobility scooter and then built out a pipe stand to brace my bike's rear wheel against the motor's shaft. Next, I wired the motor to a homemade charging circuit to regulate output and connected that to a deep-cycle battery. A common inverter then turned the battery's 12 DC volts into 120 AC volts. After a mere 3 to 5 minutes of hard-core riding, I stored enough energy in the battery to fully charge my cell phone (with juice to spare).

Here's a version of a bicycle generator that you can make work for you:

MATERIALS

- 4x4 wood posts
- An old bicycle
- 2-by-12-inch- (5-by-30-cm-) heavy plank
- Screws
- Drive belt (from an auto-parts store)
- Metal brackets
- 12-volt motor
- Heavy-duty electrical wire
- 12-volt car battery
- DC-to-AC power inverter
- Diode

BUILD

STEP 1 Measure and cut two 4x4 posts of equal length, tall enough to keep the back wheel of your bicycle at least a couple of inches off the ground when its axle is rested on top of the posts. Cut grooves into the posts for the axle to rest on.

STEP 2 Fasten the posts onto the plank with screws through their undersides, leaving enough space between them to accommodate the bike's frame.

STEP 3 Remove the bicycle's back tire, and replace it with the drive belt. Set the bike's rear axle on top of the posts and secure it in place with metal brackets.

STEP 4 Attach the 12-volt motor to the plank (add brackets if the motor has none), in line with the bicycle's rear wheel, and wrap the drive belt around the motor's axle. Make sure the drive belt has no slack.

STEP 5 Attach electrical wire between the negative terminal of the motor and the 12-volt battery, then connect the battery to the inverter's negative terminal.

STEP 6 Run wire from the motor's positive terminal to the anode end of the diode. Connect another length of wire from the diode's cathode end to the battery's positive terminal, and then wire the battery to the inverter's positive terminal.

STEP 7 Plug in your appliance to the inverter, climb on your bike, and pedal away.

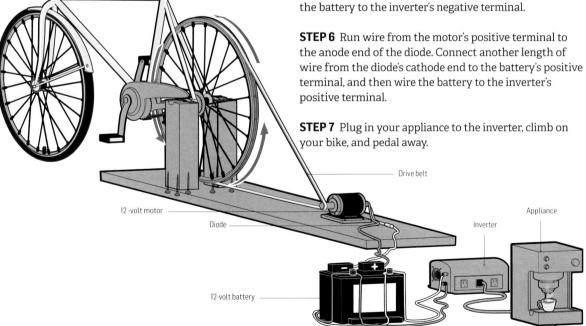

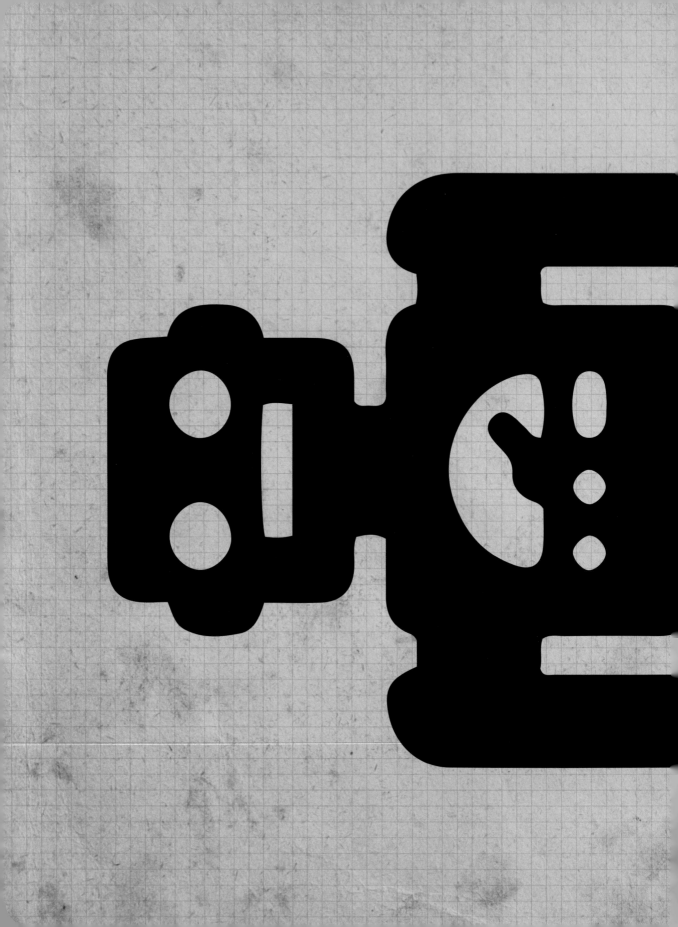

ROBOTS & BEYOND

173 MEET YOUR ROBOT MINIONS

Computer numerical control (CNC) is a fancy-pants phrase for robot tools automatically operated by computers. Theoretically, pretty much any tool can be adapted for CNC. Technically, 3D printers and other additive manufacturing systems are CNC tools as well, but in daily use the term "CNC" is applied mostly to cutting or other subtractive processes that form a shape by removing stock from a blank.

PLASMA CUTTERS These machines use a high-voltage electrical field to ionize a stream of cutting gas (usually just air sucked in from the room) into a superheated plasma that can rapidly slice through a solid steel plate. Much less expensive than metal-cutting lasers, plasma cutters work just as well for many applications.

ROUTERS In routers, rotating cutters remove material and (generally) move the cutting head while holding the stock in one place. The distinction between a CNC router and a CNC mill can get a bit confusing. Both use rotating cutters, but routing is mostly used on wood, plastic, some types of aluminum, and other so-called soft materials. Usually all cuts are completed at relatively high speeds and in a single pass. Both routers and mills are limited when it comes to inside corners, which cannot be cut finer than the radius of the cutting tool.

MILLS When you're cutting the hard stuff—primarily metals like aluminum, steel, and titanium—go with a mill. Often (but not always), a CNC milling operation involves holding the cutter stationary while the stock is moved against it. Very small desktop CNC mills are available with footprints down to 1-foot (30-cm) square, and these are especially handy for printed circuit board (PCB) milling, in which the copper on a piece of plated phenolic board is selectively cut away, leaving the circuit traces. This can be a handy way of prototyping electronics without masking or chemical etching, and because very little power is required the tools can be small, lightweight, and inexpensive.

LASERCUTTERS In a badass display of physics, lasercutters use light energy to vaporize or burn materials away from the cutting path. Lasercutting produces smooth edges on acrylic and some other plastics, and notable burned edges on wood. Lower-cost lasercutters usually can't be used on chlorine-containing plastics like PVC (because doing so releases toxic, corrosive gas) or on metals (they tend to reflect the beam). Lasercutters also allow for decorative raster effects on part surfaces, etching or burning them lightly to produce indelible marks. Unlike routers and mills, they cannot produce accurate halfway or other partial-depth cuts. On the other hand, they can cut very sharp inside corners.

WATER-JET CUTTERS These machines use a stream of very high-pressure water, often mixed with an abrasive powder, to rapidly erode and wash away material along a narrow cutting channel. Water-jet cutters are heavy, expensive, and relatively rare compared to other CNC tools, but they can do amazing things like cutting intricate, detailed shapes in thick, hard, nonmetallic materials like concrete, stone, ceramic, and even bulletproof glass.

174
SCOPE OUT THE SOFTWARE

The CNC software tool chain consists of three distinct "layers" between idea and finished part. Today, these functions are increasingly being integrated together in all-in-one programs, but it's still very helpful to understand what's going on under the hood.

CAD (COMPUTER-AIDED DESIGN) These programs are used to draw the part you want to make in two or three dimensions. Depending on your software, your CAD file may be as simple as a raster image showing a picture you want to laser-etch on a part's surface, or it may be as complex as a 3D model with multiple materials and specific instructions and parameters that change how it would be assembled depending on what tool is used to create it. CAD software, especially for 3D objects, used to be very expensive, but no more—nowadays there are a large number of free programs available online, even websites with full-featured CAD packages that run right in your browser window.

CAM (COMPUTER-AIDED MANUFACTURING) This type of software analyzes your digital model and adapts it to construction on some particular CNC tool. The CAM software figures out the exact sequence of movements, cuts, and other operations required to produce the shape contained in your CAD file. Very often these operations are represented in more-or-less standard industry formats called G-code or M-code.

CLIENT This program actually controls the tool's operation in real time. It provides basic but essential functions like start and stop, as well as various alignment, adjustment, and setting options. Depending on your particular setup, it may run on your PC or on the tool's built-in electronics.

175
GET ACCESS TO HIGH-END TOOLS

Large, expensive CNC tools have been used industrially since the 1960s, but the 21st century has seen access to these processes rapidly trickling down to the public. Even today, however, a useful CNC tool is at least a US$300 investment. And unless you're going to be using it a lot, it may not be worth it. Fortunately, there are lots of ways to get access to one.

BUILD ONE It's not outrageously difficult to put together your own CNC machine, especially if it's a router, PCB mill, or hot-wire cutter designed for working relatively soft, lightweight materials. It's a complex project, but lots of information is available online; cnczone.com is a great place to start.

BUY ONE If you need or want a CNC machine of your own, PCB mills are probably the least expensive place to start, followed by 3D printers and then CNC routers. The expensive part of a CNC machine is usually the so-called Cartesian robot, not the toolhead, and adapting one robot to use multiple tools—for example, both a router and a plasma cutter—is usually not that hard to do. Of course, a robot designed to support a 50-pound (22.75-kg) sheet of 4-by-8-foot (1.25-by-2.5-m) plywood may fold up like a used Kleenex under a 500-pound (227-kg) sheet of steel of the same size and shape, so make sure you know what you're doing before undertaking such a conversion.

SHARE ONE If there's a hackerspace, makerspace, or commercial tool-sharing collective near you, chances are it's got at least one CNC tool available for members to use. Dues, training requirements, and material and usage fees will naturally vary from one organization to another, and with the particular tool in question, so make sure you get all the details up front and run the numbers before signing up.

HIRE ONE By far the most practical option for most first-time and occasional CNC users is to leave the actual owning, maintaining, and operating of the tool to somebody else. CNC tool owners all over the world are competing online for your business and are usually happy to help walk you through the process of designing your first CAD file and uploading it to their equipment remotely.

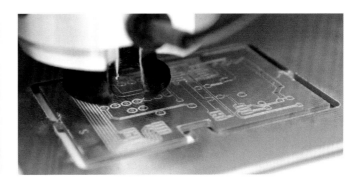

176 UNDERSTAND 3D PRINTING

A 3D printer is a manufacturing robot designed to fabricate objects, layer by layer, from digital files. Might sound like science fiction, but 3D printers have been around for more than 30 years, mostly used by engineers, industrial designers, and architects for rapid prototyping. In fact, the term is a broad label for an entire category of machines. Each type suits a particular range of materials, precision of resolution, and means of burning, melting, fusing, pressing, gluing, or stacking horizontal layers.

As you get to know the nature of the available materials and how you can push the machines to produce what you need, you'll find having a 3D printer handy on any project. You can print rare replacement parts for a classic car, for instance; build a working prototype of a mechanical device; conceptualize complex artwork; or even develop affordable prosthetics. You're only limited by your imagination—and the amount of time you have to tune and mod your printer and printer files.

FUSED-FILAMENT FABRICATION (FFF) The most common type of desktop 3D-printing technology, also known by its trademarked name Fused Deposition Modeling (FDM). FFF machines create objects by "drawing" them in melted plastic, extruded through the nozzle of a guided toolhead. For each layer, the nozzle traces an object's outer shell and then follows an infill pattern to fill in the interior of the part at a user-selected density. By depositing layer after layer, each new slab fusing to the one underneath, a complete physical object takes shape.

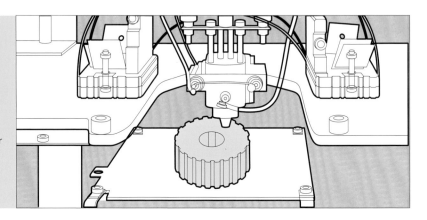

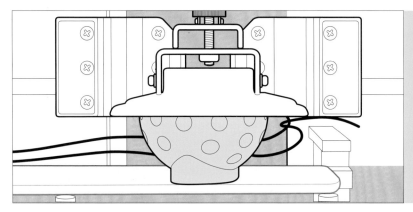

STEREOLITHOGRAPHY (SLA) Unlike the FFF process, where a toolhead travels across the build platform, squirting out the object in melted plastic, SLA machines use a mirror to selectively cure the surface of a photoreactive resin. After every laser pass, the object is slowly pulled up out of the vat of resin. Recently, a number of new desktop SLA printers have come to market, offering finer-resolution printing but with the tradeoff of longer print times, fragile parts, and expensive and sometimes toxic printing materials.

"Makers love to trade skills. So what are you good at? Chances are, the owner of a 3D printer would swap woodworking chops or welding advice for access to their machine."

177 FIND A 3D PRINTER NEAR YOU

Thanks to the many new companies now producing well-built, affordable 3D printers, you may have opportunities to try out a 3D printer in your area before considering whether you'd like to invest in one of your own. Besides hackerspaces, ere are a few places to look:

LIBRARIES As part of an effort to bring the local communities back into their public libraries, locations all over the country are starting to set up dedicated spaces for hands-on projects, many of which include 3D printers.

SCHOOLS AND UNIVERSITIES Educational institutions have been exploring 3D printing for decades to satisfy curriculum needs for engineering, architecture, mathematics, and design. Depending on when your local schools purchased equipment, you might see anything from older industrial-grade models to rooms full of more affordable, modern desktop models.

3D-PRINTING STORES AND SERVICES Physical 3D-printing stores are mostly in major cities, but there are lots of new locations opening to serve the growing demand for 3D-printed objects. There are also a growing number of services that will print your projects for you. These services can be expensive and take a few weeks, but they offer the ability to print in a wider variety of materials—from nylon to precious metals to ceramics—on industrial-grade 3D printers.

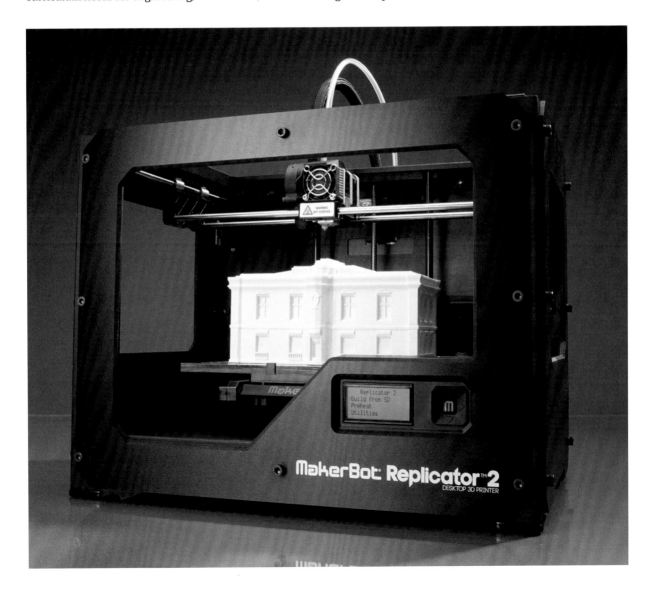

178 PREPARE TO PRINT A 3D MODEL

WORK IT GOOD

The first step to printing a 3D model is to prepare the digital file, which tells the printer what exactly to print.

STEP 1 Pick an object to print. You'll need a 3D digital model, preferably an .stl (stereolithography) file with solid, watertight geometry. Your design can be created using a mechanically inclined computer-aided design (CAD) software package, a 3D digital art and animation tool, or even captured directly with a 3D scanner. (You may need to clean up a 3D scan before it's ready for printing.) The .stl file format was created decades ago and can be exported from most CAD and computer graphics software packages.

There are also a number of online 3D model repositories where designers and artists share their latest work: Thingiverse, YouMagine, GrabCAD, and 3D Warehouse, among others. Browse the sites by object name, category, or designer, and check to see that the object you have selected offers a downloadable model that matches your needs. At right are some examples of very cool things you could find plans for online or develop plans for yourself—from DIY musical instruments to functioning tools, and from a complete chess set to architectural models and actual bone replacements.

STEP 2 Import the file into computer-aided manufacturing (CAM) software. CAM software helps you translate your digital design into real-world instructions that a 3D printer can follow. Check with the manufacture of your 3D-printer and others who use the same model for recommendations on which package (and preset configurations) best suit your printer model.

STEP 3 Preview your design. When you import the .stl, an image of the model appears in the window that represents the build platform (the area where the printed part will be produced) for your specified printer. You may need to scale down your digital model to match the printed dimensions. (If your model is too small, you might need to convert your digital design from inches to millimeters—scale your model up 25.4 times to convert it.) Other CAM preview features allow you to reposition, rotate, mirror, and duplicate your design until the virtual build platform matches what you intend your printer to produce. Once you move the scaled-down version into the center of the platform and it fits within the bounds of your printer, it's ready for "slicing."

STEP 4 It's time to lean on the core power of your CAM package. Slicing tools (either included in the CAM tool or available as a separate application) are translation utilities that analyze the virtual geometry of your models and create a set of machine-specific instructions your printer will follow to produce the physical object. The term "slicing" refers to dividing the model into horizontal layers first before working out the path the toolhead will follow to deposit each layer of plastic. Before slicing, configure the job-specific print settings that the slicing tools will follow to produce the object.

STEP 5 Look closely at your file to see if there are any challenges for your slicer to overcome—say, the model isn't completely flat on the bottom (a problem for a stable print), or there are no lower features to support the top of the object (referred to as "overhang"). To facilitate this print, activate "raft" and "automatic support" settings in the slicer. A raft is a low, wide pedestal for your design to sit on top of—to provide a stronger base for the bottom of your model. (A similar feature is a brim, which rings the base of the model like a hat brim.) Automatic support options generate breakaway structures to support features higher up in the model that would overhang empty space otherwise. Both rafts and support features are designed to be easy to remove from the original design after the printing process. Most slicer packages offer these options.

STEP 6 Consider other options you may want to adjust from job to job. There's layer height, which determines how thin to slice each vertical layer. Thinner layers mean softer, less blocky curves up the side of your object, but multiply the amount of time necessary to produce the object. Then there's fill density, which determines how tightly to pack the insides of solid objects. Higher density produces stronger, heavier parts but also adds to printing time and amount of material required.

STEP 7 Slicing the file may take a few seconds or a few minutes, depending on the complexity of the job and the slicer you've selected. When it's done, you're ready to print your object! Depending on what specific 3D printer you're using, hitting print varies. Refer to your printer's instructions for specifics. **Pro tip:** Most FFF 3D printers will leave a tiny seam up one side of the model based on where the toolhead moves vertically to create the next layer. After finding where your machine adds a seam, you may want to rotate the design before printing the next time, to ensure that the seam doesn't mar a significant detail on your model (like the face on a 3D portrait).

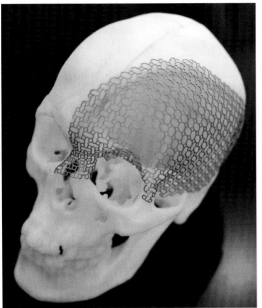

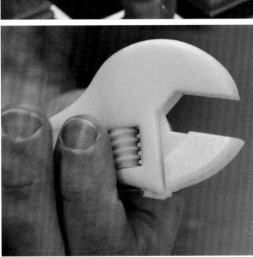

HACKETT SAYS

179 ENTER THE NEW AGE OF REPLICATORS

First hearing about 3D printing was exciting and kind of validating—finally, the future. Because as the 21st century aged, the future looked to be a massive bait and switch. We were promised wonders—jet packs, virtual reality, Soylent Green. Instead, we got more efficiently targeted ads. But a real-deal replicator? That more than made up for the disappearance of privacy.

The promise was amazing, but the actuality was lame and tacky. Instead of carbon-knitting nano wonders, we just had more crap made out of plastic. As usual, the disconnect between marketing and reality is incredible.

But through the generosity of a certain Brooklyn-based 3D-printer concern, my shop has a tiny toehold in the 21st century. I can now make some things that the world urgently needs, like the tiny model of my head shown here, and prototype slightly slower than carving soap.

Still, I think the prospect is promising—the future has moved slightly closer, and if I shake off marketing hype and squint, I see the grand disappointment of my youth, the clunky command-line PC that taught a generation to code. What we have now is the seeds planted for the next couple of generations: Today's 3D printers are the Commodore 64s and Sinclairs for those alive 30 years from now—garbage, really, but garbage that spurred a revolution.

180 FOCUS ON PLASTICS

At the chemical level, plastics are essentially long chains of polymers whose monomer building blocks are derived mostly from fossil fuels through a series of heat and chemical reactions. Plastics come in a huge variety, each with their own particular characteristics—from soft and flexible to hard and brittle. The shapeshifters of the synthetic physical world, plastics' chemical makeups allow them to take on nearly any color or shape.

NEED SUPERSTRONG CEMENT? SHREDDED PLASTIC (TRY A TARP) WILL REINFORCE CEMENT LIKE LIGHTWEIGHT REBAR.

MAKE YOUR OWN SIMPLE BIOPLASTIC WITH VINEGAR, WATER, STARCH, GLYCERIN, AND A LITTLE BIT OF HEAT.

PLASTIC LASTS FOR AT LEAST 500 YEARS. USE IT IN YOUR PROJECTS TO KEEP IT OUT OF THE ENVIRONMENT.

WHILE MOST PLASTICS ARE MADE FROM RAW MATERIALS TAKEN FROM FOSSIL FUELS AND ARE NON-BIODEGRADABLE, SOME BIOPLASTICS WILL BIODEGRADE OVER TIME; THESE GET THEIR BUILDING BLOCKS FROM PLANT SOURCES LIKE CORN, SOY, OR HEMP.

DON'T BURN PLASTICS. IN GENERAL, MANY PLASTICS GIVE OFF DIOXINS—HIGHLY TOXIC CHEMICALS KNOWN TO HAVE NEGATIVE EFFECTS ON HEALTH—AND NASTY HYDROCHLORIC ACID WHEN BURNT.

SOME PLASTICS ARE CHEMICALLY INERT. RULE OF THUMB: IF IT HELD A BEVERAGE, IT'S SAFE TO HOLD ACID OR GASOLINE.

WE EACH TOSS 185 POUNDS (84 KG) OF PLASTICS A YEAR. TRY FORAGING SOME FOR A PROJECT INSTEAD.

PLASTIC IS THE PAPER OF RAPID PROTOTYPING. A NUMBER OF DIFFERENT TYPES COME IN SPOOLS OF COLORED FILAMENT AND ARE USED FOR 3D PRINTING. IF YOU CAN THINK IT UP AND CONJURE IT WITH MODELING SOFTWARE, YOU CAN PRINT IT.

TURN FLIMSY PLASTIC BAGS INTO A DURABLE MATERIAL. Snip the handles and the bottom off a bag. If there's writing on it, flip it inside out. Fold the bag into an eight-ply square and place it between two sheets of paper (wax paper, parchment paper, or printer paper) on an ironing board. Turn off the steam on your iron and set the heat to the rayon setting. In a well-ventilated space, iron for 10 to 20 seconds on one side, moving the iron constantly. Flip and repeat. Sew the resulting material into reusable bags, clothing, or a sail for your plastic-bottle boat.

PLASTICS ARE USUALLY SUCH POOR CONDUCTORS THAT THEY'RE OFTEN USED TO INSULATE WIRES AND ELECTRICAL CABLES.

IF IT KEPT LIQUID IN, IT CAN KEEP LIQUID OUT: EMPTY BOTTLES ARE THE BOAT OF YOUR DREAMS, JUST A DUMPSTER DIVE AWAY. AN EMPTY ½-LITER WATER BOTTLE WILL KEEP A POUND AFLOAT. FIGURE OUT YOUR CARGO, CONNECT BOTTLES WITH DUCT TAPE, AND ADD A 50 PERCENT MARGIN TO ACCOUNT FOR LEAKS AND EXTRA WEIGHT. SOON: A DURABLE, FLOATING MONUMENT TO MAKER RECYCLING.

181 PICK A MICROCONTROLLER

Microcontrollers are tiny, cheap computers. But unlike the computer on your desk or in your pocket, they don't come with a screen or a mouse. You get to choose what inputs and outputs to add to create new devices. Make a doggy door that unlocks when the correct pet comes near, a coffeemaker controlled by an alarm clock, a robot that sorts candy by color, or anything else you want to automate.

RASPBERRY PI Almost a full computer the size of a deck of cards. Has video, audio, and Internet built in, but it lacks the analog inputs that most other boards have. Great for giving your robot vision or making a home media computer.

ARDUINO UNO The poster child of microcontrollers, it's by far the most popular, which means it has tons of community support and accessories. Wimpy compared to computers, it can still do a lot.

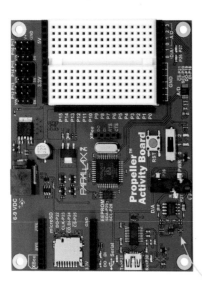

ARDUINO YÚN Full Arduino, like the Uno, but has an entire extra computer packed in that controls WiFi, Ethernet, and a microSD card, freeing the Arduino to do more important things.

PARALLAX PROPELLER Eight low-power processors on one chip, all running independently. Great for robots and other projects that you want to design to do many things at the same time.

ADAFRUIT FLORA Small, Arduino-compatible, and designed to be worn. All the pros of an Arduino Uno but made for sewn projects. Has a connector for rechargeable batteries, too.

TI LAUNCHPAD Alternative to the Arduino ecosystem, made by Texas Instruments. Many affordable boards with different capabilities.

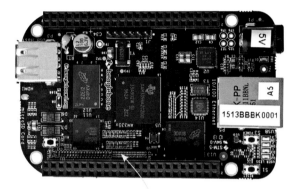

BEAGLEBONE BLACK A small, full-featured computer similar to the Raspberry Pi. Has tons of ports for sensors and other inputs and outputs.

PARALLAX BASIC STAMP Similar in capabilities to the Arduino but has been around for decades. Popular in education and unique for its simple PBASIC programming language.

INTEL GALILEO Essentially a full-blown computer, based on Intel architecture, with support for Windows, Mac, and Linux. Can use both Arduino shields and code.

182 ACCESSORIZE YOUR MICROCONTROLLER

The great thing about microcontrollers is that you start with the bare bones and add what you need. But it also means you usually need a few extra things to get started. Check the specifics of your board to see exactly what tools are required.

COMPUTER A "real" computer of some kind is typically needed to write programs or transfer them to the microcontroller. Any computer made in the last five years should be more than enough.

PROGRAMMING CABLE This is usually just a USB cable, but some boards do require a special cable or a memory card to move programs over.

POWER Some can get power from a USB cable; others require an external power source.

MEMORY CARD Many boards use SD cards for storage. Check the specs for the right size, shape, and speed.

183 MAKE A CASE FOR PROTOTYPING

Once you discover microcontroller basics, it's natural to start tinkering with different ideas. Create a prototyping box to help keep the experimental bits and bobs contained and the cables where you can find them.

STEP 1 Find a plastic case with a closable lid and adjustable dividers. A single-layer box with 2x8 compartments is good for basic projects, but bigger ones mean more stuff. **Pro tip:** Craft and fishing supply stores will give you a lot of options.

STEP 2 Attach the microcontroller and a solderless breadboard next to each other on the inside of the lid. Solderless breadboards have foam tape on the bottom to hold them in place, and you can use hot glue or machine screws to hold the microcontroller. Be sure to allow clearance along the edge of the microcontroller for power and programming cables. You might have to cut holes in the case to give them room.

STEP 3 Use a sharp craft knife or a rotary tool to cut down the dividers that intersect the breadboard and microcontroller so the lid will close.

Arrange the rest of the dividers to hold your stuff, starting with the biggest pieces and working down to the smallest.

STEP 4 Fill the box with your various bits and bobs. Programming cables, power supplies, jumper wires, LEDs, various resistors (220-ohm, 560-ohm, and 10k-ohm are a good start), micro servos, push buttons, piezo buzzers, potentiometers, photoresistors, temperature sensors, and whatever else you want to experiment with.
Pro tip: Throw in a couple of rubber bands and binder clips—they always come in handy when your actual hands are busy tinkering.

184 USE PROGRAMMING TO BLINK AN LED

The best way to see if you have your microcontroller set up correctly is to make an LED blink. It's also the first step to building more complex projects. Add a button to it and you'll have both elements that make microcontrollers so handy: inputs and outputs. Inputs feed information into the little computer and outputs react to it.

STEP 1 Wire an LED to your microcontroller. Some boards come with a blinkable LED built in, but that's no fun. Use a solderless breadboard and some solid-core hookup wire to connect an LED and a 220-ohm resistor to the microcontroller. Connect the short pin of the LED to ground on the microcontroller. The resistor goes between the long pin of the LED and any digital pin on the microcontroller. **Pro tip:** The 220-ohm resistor is not the perfect choice, but it'll work with most combinations of LEDs and microcontroller voltages. If the LED is very dim, try removing the resistor.

STEP 2 Install the board's programming tool, available on the manufacturer's website. Open it up and write code to blink the LED. For most microcontrollers, the program needs to follow these steps: Set the LED pin to Output mode, turn the pin on and wait a bit, turn the pin off and wait a bit. The code for Arduino and compatible microcontrollers is available at popsci.com/bigbookofmakerskills.

STEP 3 Connect the microcontroller to your computer with the recommended cable (or memory card) and upload the code. There's usually a tool in the programming environment that will help you do this. Microcontrollers automatically run their program when reset, so if all is well the LED will start blinking a few seconds after it's done uploading. If not, check your circuit and code for mistakes. If you're stuck, most microcontrollers have tutorials for how to blink an LED.

STEP 4 Add an input to the project by connecting a push button. Connect one side of the button to +5 volts (or +3.3 volts on boards that use 3.3 volts). The other side of the switch connects both to a digital pin and to a 10k-ohm resistor that goes to ground.

STEP 5 Have the microcontroller read the button state. The process looks something like this: Set the button's pin to Input mode, see if the pin is high (pressed) or low (not pressed). If it's pressed, do something. If it's not, do something else (or nothing at all). The Arduino code, modified to blink the LED when the button is pressed, is available at popsci.com/bigbookofmakerskills.

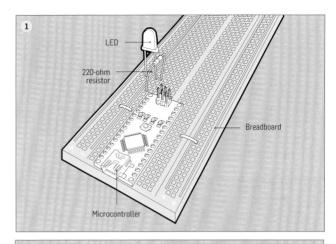

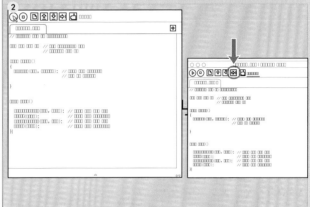

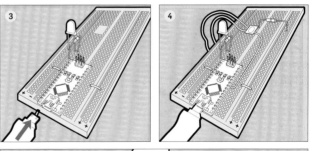

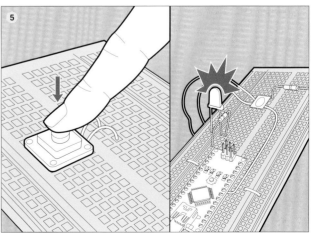

185 PICK AN INPUT OR OUTPUT

A microcontroller's power doesn't come from its wimpy CPU but from the unlimited combination of things you can connect to it. Inputs like buttons, switches, and sensors provide the microcontroller with information, and outputs like lights, motors, and actuators act on it. Here are a few to spark your imagination.

BRAINWAVE SENSOR Give your project a mind of your own by listening to your brainwaves. It won't read specific thoughts, but it can react to concentration, mood, tension, and other general brain activity. It can require a powerful microcontroller to correctly interpret brainwave data.

ULTRASONIC RANGE FINDER Keep your robot from smashing into things using echolocation. This sensor measures distance by sending out an ultrasonic pulse and measuring how long it takes to hear the echo. Accurate to 1 inch (2.5 cm) at 20 feet (6 m) or more.

AC POWER SWITCH Like a power strip that you can turn on and off with a microcontroller. Plug in small appliances, lights, pumps, and other things that work on household current for easy home automation. Note that some appliances don't react well to very quick switching.

THERMOELECTRIC ELEMENT Apply power to these and one side gets cold while the other gets hot. (Be sure to add a heatsink to the hot side or it can burn up!) If you reverse it and expose each side to a different temperature, it'll generate a small amount of power.

SERVOMOTOR A high-torque motor that can be rotated to precise angles, typically limited to 180 degrees (or less) of rotation, but that's enough for most robot limbs. It's easy to attach things to a servomotor using the various "horns" that connect securely to the servo axle.

PIEZO ELEMENT Apply power and it will beep like the warning of a smoke detector. Pulsing the power can produce musical tones. Or you can reverse the process: Flexing the crystal generates power, making it into a simple vibration sensor.

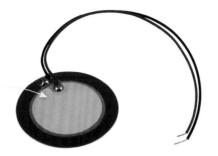

GPS A global positioning module will not only give your project's location within a few feet, it can also provide the time accurate to within a hundred-millionth of a second. Add an external antenna to increase accuracy and reliability.

ADDRESSABLE LEDS Let you easily control the brightness and color of dozens (or hundreds) of lights at once. They come in squares, strips, rings, or any other shape you need to show text, images, or mood.

FLEX SENSOR Measures how far something bends. Often used in gloves to read the position of fingers, it can be used anywhere you want to know how much something bends. They require some careful mounting to bend without breaking.

AIR SENSOR Just because you can't see it or smell it doesn't mean it's not there. With an air sensor, you can detect the presence of carbon dioxide, oxygen, alcohol, methane, particulate pollution, and many other airborne substances.

LINEAR MOTOR When you're tired of motors spinning things in a circle, you can use one of these to move them in a straight line. Low speed and high power mean they can push or pull lots of weight with high accuracy.

RFID Radio frequency identification tags are the size of a coin, each with a unique ID. An appropriate tag reader will remotely sense the presence (or absence) of a tag from 2 inches (5 cm) to 2 feet (60 cm) away. Many mobile phones can also work as a tag.

186 TAKE A SUPERQUICK PROGRAMMING PRIMER

To program a computer you start with a task, then break it into smaller and smaller steps until you end up with actions simple enough for a computer to do. Then you explain those actions in a way that a computer can understand. That means speaking a computer's language. There are dozens of popular computer languages out there, but they're not full languages like Mandarin or Swahili. They're simple English with very special rules. The rules are different for each language, but they all have common concepts.

SPECIAL PUNCTUATION Trust me, minding your dashes and brackets carries more weight in programming than in grade-school English class. Forgetting a parenthesis or using a comma instead of a period can stop a program dead. Here are some common examples of special punctuation:

- **Comments** These are used as a chance to inject some plain English into your code and are ignored by the computer. They often start with # or // and end at the end of the line, or begin with a /* and end with */.
- **Code Blocks** Handy for grouping a bunch of code together as a single unit. This is usually done with curly brackets, but some languages group code by how far the line is indented.
- **End of Line** This is commonly indicated by a semicolon, while some languages are smart enough that you can just hit return. Computers are not very smart about knowing where the end of a line of code is.
- **VARIABLES** The labeled boxes of coding. You can put a single thing in the box; for example, a number or word. Then you can do things with the boxes, like add or compare their contents without knowing exactly what's inside. And you can give them descriptive names so you can tell what kind of thing they're supposed to hold. **Note:** They don't start with a number, and they can't have spaces.

CONDITIONALS These let the computer make decisions by comparing things, usually variables. Here's an example:

```
if (x > 10) {
    // Do something.
} else {
    // Do something else.
}
```

In plain English, this reads "If the value of variable X is greater than 10, then do whatever is in the first block of code. Otherwise, do the second block of code." Conditionals sometimes use borrowed and improvised logical symbols, like != for "not equal to" or >= for "greater than or equal to."

LOOPS Use a loop to make computers do the same thing over and over without complaining. They're like conditionals with something extra going on:

```
for (x=0; x < 10; x=x+1) {
    // Do something.
}
```

In plain English: "Set X to zero. If X is less than 10, do the block of code. When the block is complete add one to X. If X is less than 10, do the block of code..." etc., adding one to X each time through. When X is no longer less than 10, the loop stops and the computer moves on.

FUNCTIONS If you have a bit of superuseful code, you can split it into a function that you can reuse without copying every time. Here's an example in the JavaScript language:

```
// A small function to return the cube of a number.
function cube(num){
    var output = num * num * num;
    return output;
}

alert("The cube of 12 is " + cube(12));   // 12 cubed is 1728
alert("The cube of 42 is " + cube(42));   // 42 cubed is 74088
```

Here, we send a number (num) to the function, which cubes the number, returns the answer (output) to where it came from, and displays it on-screen. A basic JavaScript program that combines these concepts is easy to find online.

187 BUILD AN ADD-ON BOARD

Most microcontroller boards have sockets to quickly attach or remove extra components, making it easy to add features or move the board to other projects. Often called shields or capes, these self-contained bits of circuitry are a great way to make your temporary circuits more permanent and keep your microcontrollers from becoming entombed in an old project.

STEP 1 Use wire cutters to cut breakaway headers to match the pins on the microcontroller. For an Arduino Uno, it's two groups of eight pins and two groups of six. If your board has male pins (like the Raspberry Pi), use female headers. Some microcontrollers don't come with any pins so you'll have to add your own.

STEP 2 Socket the headers into place on the microcontroller and place a piece of double-sided perfboard on top. Line it up so the header pins poke through the evenly spaced holes. Full-size Arduino boards have unusual spacing on eight of the pins—you'll need pliers to bend doglegs in the pins so they line up.

STEP 3 Cut the perfboard to size. Use a straight edge and a utility knife to score along a row of holes on both sides. After scoring several times, just snap it along the line. Use a file or sandpaper to clean up any sharp edges. The center layer of perfboard is fiberglass, so use eye protection and a mask when cutting or sanding it. **Pro tip:** Don't use your good tools, since fiberglass will quickly blunt them.

STEP 4 Place the perfboard back on the pins and solder them in place. Soldering the pins to the board when it's mounted on the microcontroller will keep the pins properly aligned.

STEP 5 Since you've just covered up most of the labels on the microcontroller board, use a fine-tipped permanent marker to labels the pins on the top of your perfboard. **Pro tip:** Add the project name to the board, so when you find it in a drawer later you can remember what it does.

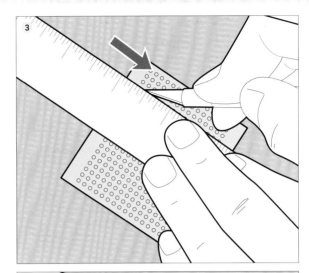

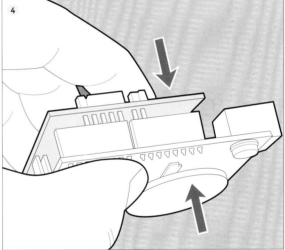

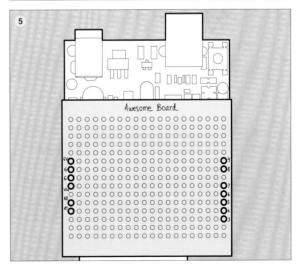

188 HAIL THE LASER-CUTTER

Albert Einstein had ideas about laser technology as far bask as 1917. And in 1997, Dr. Evil wanted to attach frickin' lasers to the heads of sharks. Today, lasers fit in small packages and many businesses, schools, and shops have their own lasercutters.

The laser is housed in a ventilated box over a cutting bed. Computer numeric control (CNC) moves the focused beam using a digital grid of precise points along two axes. Think of the laser gantry being able to move north and south, and east and west, when instructed by the digital design.

Although inexpensive to use, lasercutters are not cheap to own—currently it costs between US$3,000 and US$10,000 for basic versions. Expect to take a class on best practices before you can schedule time on a machine at a junior college or makerspace. Additionally, each manufacturer has its own operating system and software. Changing brands means learning a new tool.

189 LASERCUT THE RIGHT MATERIALS

Lasercutters have limitations when selecting materials to cut or etch. Materials that emit harmful gases when heated or melted are prohibited. Any material that could reflect the laser beam into the focusing lens will cause permanent damage. Human body parts are always a bad idea (check YouTube for cringe-worthy examples). The facility owner and laser manufacturer can provide a list of approved materials, but here's a list of possible options:

ACRYLIC Leaves a smooth, finished edge that's immediately ready for use or assembly.

WOOD Makes for precise, smooth cuts; edges have a special burnt appearance.

CLOTH Adjust the intensity to limit charring.

LEATHER While etchings are beautiful and detailed, the process stinks—warn your shopmates and be sure to ventilate.

PAPER Very smooth cut lines—ornate patterns are mesmerizing.

CLOSED-CELL FOAM Results in a perfect finished edge.

APPROVED RUBBER Note that thicker cuts are rougher.

APPROVED FOOD Etch fruits and veggie skins for a novelty snack—or make steak business cards.

GLASS Etching on glass requires a rotating jig (a rotary).

ANODIZED ALUMINUM Creates a nice contrast in any color and boasts better detail than the usual engraving.

190 MAKE YOUR FIRST LASERCUT DESIGN

Once you've located a lasercutter and gone through the necessary classes, all that remains between you and etching your own custom headstone is the design itself. Use Adobe Illustrator or CorelDRAW to handle your graphics. Support for these programs is easy to find. Don't overthink it.

STEP 1 Choose a known good file or download one from the Internet. Save it to a thumb drive. Run this file first so you can make sure the machine is working properly.

STEP 2 Start your graphics program and open up a blank document. Draw a design or import a graphic and convert it to vector art. **Pro tip:** For photographs, sharp contrasts are better than subtle shades. Convert the image to a line drawing. Then, select the correct thickness. When you are done, save it as an Adobe Illustrator file, CorelDRAW file, or PDF.

STEP 3 Confirm that your material is approved for lasercutting and that it's the right size to fit the cutting bed. Smaller pieces of material will need to be fixed or weighted so they don't move. Have plenty of practice material handy.

STEP 4 Import your job file. Select the appropriate settings. Place your material in the lasercutter. Tell the laser where to begin—or "establish home." Your files will cut or etch in a matter of minutes.

191 TROUBLESHOOT A LASERCUTTING JOB

Most problems encountered while lasercutting are common, easily addressed, and fall into three categories: files, formatting, and importing; tuning settings to suit materials and design; and operational issues with the lasercutter.

FILES, FORMATTING, AND IMPORTING

- Job does not show up: Select "job type," and make sure Resolution is set to "All."
- Laser etches instead of cuts, reads "job done," and won't start: Your file is not vector art.
- Job runs but etches instead of cuts, or makes no visible marks: Your line weight is incorrect.
- Unwanted cuts or patterns appear: Objects are hidden on other layers of your graphic file.
- File not accepted, art is too small, or odd placement: Your graphic is sized to the wrong specifications and needs to be adjusted in your chosen software.

TUNING SETTINGS TO SUIT MATERIALS AND DESIGN

- Machine moves through file with no result: Your laser is not focused.
- Flaring or flaming: Speed is set too slow or power too high.
- Does not cut cleanly: Power is set too low or speed too fast.
- Not enough contrast on etching: Pulses Per Inch are set too low.

OPERATIONAL ISSUES WITH THE LASERCUTTER

- Does not cut: The beam is diffused from a dirty lens.
- Light point does not show or is not centered: The mirrors are misaligned.
- Smoke buildup: There is a ventilation problem.
- Material moves about during run time: Your air assist is too high.
- Laser does not move and gives error code: There is a physical failure of drive belts or axis motors.

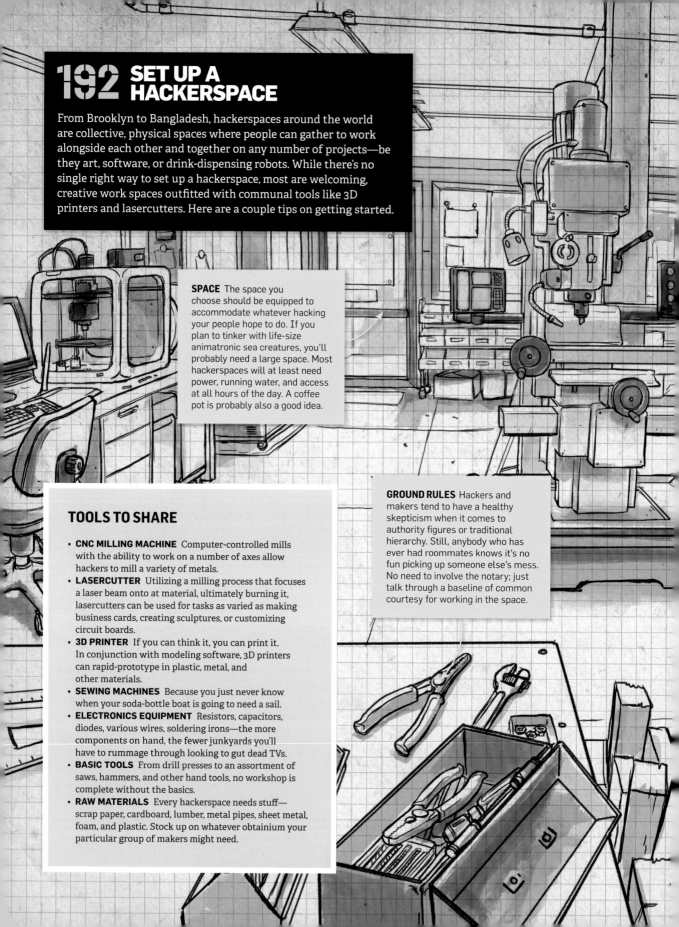

192 SET UP A HACKERSPACE

From Brooklyn to Bangladesh, hackerspaces around the world are collective, physical spaces where people can gather to work alongside each other and together on any number of projects—be they art, software, or drink-dispensing robots. While there's no single right way to set up a hackerspace, most are welcoming, creative work spaces outfitted with communal tools like 3D printers and lasercutters. Here are a couple tips on getting started.

SPACE The space you choose should be equipped to accommodate whatever hacking your people hope to do. If you plan to tinker with life-size animatronic sea creatures, you'll probably need a large space. Most hackerspaces will at least need power, running water, and access at all hours of the day. A coffee pot is probably also a good idea.

GROUND RULES Hackers and makers tend to have a healthy skepticism when it comes to authority figures or traditional hierarchy. Still, anybody who has ever had roommates knows it's no fun picking up someone else's mess. No need to involve the notary; just talk through a baseline of common courtesy for working in the space.

TOOLS TO SHARE

- **CNC MILLING MACHINE** Computer-controlled mills with the ability to work on a number of axes allow hackers to mill a variety of metals.
- **LASERCUTTER** Utilizing a milling process that focuses a laser beam onto at material, ultimately burning it, lasercutters can be used for tasks as varied as making business cards, creating sculptures, or customizing circuit boards.
- **3D PRINTER** If you can think it, you can print it. In conjunction with modeling software, 3D printers can rapid-prototype in plastic, metal, and other materials.
- **SEWING MACHINES** Because you just never know when your soda-bottle boat is going to need a sail.
- **ELECTRONICS EQUIPMENT** Resistors, capacitors, diodes, various wires, soldering irons—the more components on hand, the fewer junkyards you'll have to rummage through looking to gut dead TVs.
- **BASIC TOOLS** From drill presses to an assortment of saws, hammers, and other hand tools, no workshop is complete without the basics.
- **RAW MATERIALS** Every hackerspace needs stuff—scrap paper, cardboard, lumber, metal pipes, sheet metal, foam, and plastic. Stock up on whatever obtainium your particular group of makers might need.

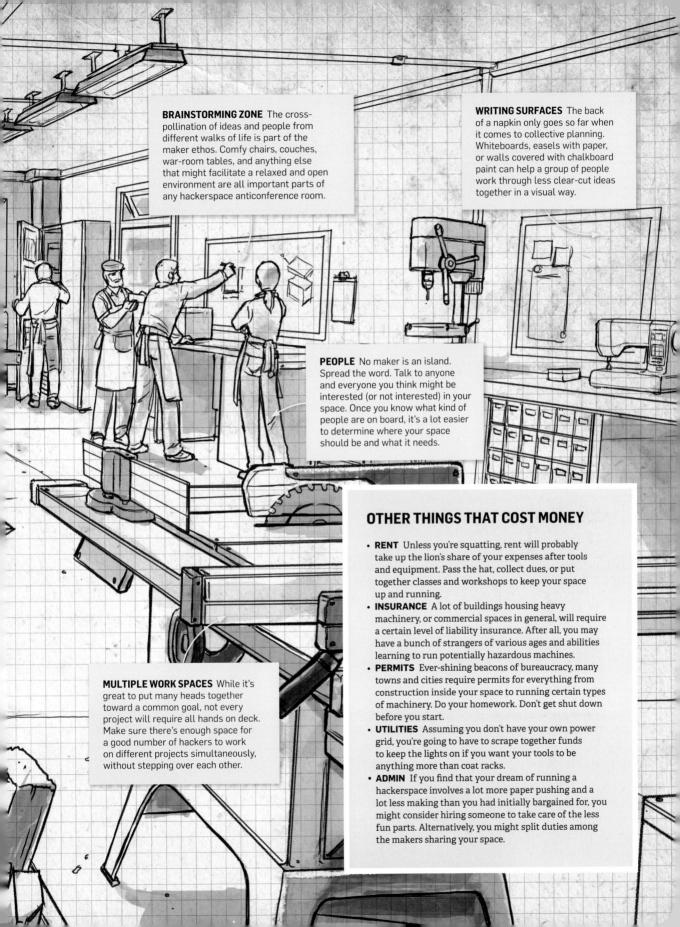

BRAINSTORMING ZONE The cross-pollination of ideas and people from different walks of life is part of the maker ethos. Comfy chairs, couches, war-room tables, and anything else that might facilitate a relaxed and open environment are all important parts of any hackerspace anticonference room.

WRITING SURFACES The back of a napkin only goes so far when it comes to collective planning. Whiteboards, easels with paper, or walls covered with chalkboard paint can help a group of people work through less clear-cut ideas together in a visual way.

PEOPLE No maker is an island. Spread the word. Talk to anyone and everyone you think might be interested (or not interested) in your space. Once you know what kind of people are on board, it's a lot easier to determine where your space should be and what it needs.

MULTIPLE WORK SPACES While it's great to put many heads together toward a common goal, not every project will require all hands on deck. Make sure there's enough space for a good number of hackers to work on different projects simultaneously, without stepping over each other.

OTHER THINGS THAT COST MONEY

- **RENT** Unless you're squatting, rent will probably take up the lion's share of your expenses after tools and equipment. Pass the hat, collect dues, or put together classes and workshops to keep your space up and running.
- **INSURANCE** A lot of buildings housing heavy machinery, or commercial spaces in general, will require a certain level of liability insurance. After all, you may have a bunch of strangers of various ages and abilities learning to run potentially hazardous machines.
- **PERMITS** Ever-shining beacons of bureaucracy, many towns and cities require permits for everything from construction inside your space to running certain types of machinery. Do your homework. Don't get shut down before you start.
- **UTILITIES** Assuming you don't have your own power grid, you're going to have to scrape together funds to keep the lights on if you want your tools to be anything more than coat racks.
- **ADMIN** If you find that your dream of running a hackerspace involves a lot more paper pushing and a lot less making than you had initially bargained for, you might consider hiring someone to take care of the less fun parts. Alternatively, you might split duties among the makers sharing your space.

193 PICK A CROWDFUNDING PLATFORM

Now that you have the tools and skills, you may find yourself making cool stuff that others would also enjoy and want. You have an idea, a project, a product, and now all you need is the funding to go big. Once upon a time, you'd have to go in search of an investor or two. But now, you can present your idea to the world online and end up with investors, each coming in at various levels of investment. Welcome to the world of crowdfunding.

There are literally hundreds of crowdfunding platforms to choose from, each with its own rules and regulations. Which platform will suit you best depends on the project you're trying to fund. A solid place to start is deciding whether your campaign will be based on equity or rewards. Will people who contribute become shareholders or just get a premium?

If what you're funding is a startup, you may want to go the equity crowdfunding route. AngelList, CircleUp, Crowdfunder, and EquityNet are a few of the most popular.

Like many makers, if what you're funding is a product, you'll want to launch a rewards-based campaign. Two of the most popular platforms among the maker community are Kickstarter and Indiegogo, each with its pluses and minuses. Kickstarter is only open to "creative projects," where a project is defined as "a finite work with a clear goal." What this means is that you can't fund, say, starting a business. You can pretty much fund anything on Indiegogo, but the community is much smaller.

Another major difference is that Kickstarter is an all-or-nothing model—if you don't hit your monetary goal by the end date of your campaign, your backers are not charged and you get nothing. With Indiegogo, you get to keep what you raised, even if you don't hit your goal. On the downside, the lack of urgency among backers could potentially limit funding.

A third major difference, at the time of this printing, is that Indiegogo is an international platform, running in hundreds of countries and regions across the globe. Campaigns can be in English, German, French, and Spanish, and contributions can be made in USD ($), CAD (C$), AUD (A$), EUR (€), or GBP (£). And while anyone is allowed to fund a Kickstarter campaign, the campaigns must be based in the United States, United Kingdom, Canada, Australia, New Zealand, or the Netherlands.

Also bear in mind that there are specialty platforms out there for specific things, like music and nonprofits. No matter what platform you decide on, be sure to read all the fine print so you understand fully what you're signing up for.

194 PUT TOGETHER A PITCH VIDEO

Very likely the most important part of your crowdfunding campaign is your pitch video, which can oftentimes double your success rate. As we all know, though, the Internet is overflowing with clever things vying for a person's attention. Here are some pointers to make your video shine among thousands.

STEP 1 Watch a bunch of pitch videos, particularly ones from successful campaigns, and note what works and doesn't work for you as a viewer.

STEP 2 Before you get started scripting your video, clearly define what you want to share. People are interested in the personal story. What can you tell them about yourself, your passions, and your knowledge? What are you proposing to do, why, and how?

STEP 3 You don't have to have a professionally produced video to draw attention, but be sure you do have solid sound and lighting so folks can actually see and hear you.

STEP 4 Identify what interesting visual elements you can include. People want to see that you're a genuine human, so show yourself, but also show sketches, proof of concept, and the final product, if you have it.

STEP 5 Be real. You don't have to be an actor or actress. Just be yourself. And be abundantly clear and transparent on what exactly you intend to do and what the funds will be applied toward.

195 BUILD COMMUNITY

Running a campaign is not just a matter of making a video, putting up the web page, and waiting to see what happens—it can be a full-time job. Folks are ready to support your great idea, but they generally want to feel like they're part of the development. These early adopters are especially interested in hearing about your process and being kept informed on the hurdles. It's like buying a backstage pass.

PROMOTE Identify which communities would be interested in your project and reach out. Get out and give talks where you can, like at conferences and meet-ups. Connect with media outlets to share why your story is interesting and important. Packaging up compelling images and video links makes their jobs easier and your story more likely to get picked up. Make a press kit.

INFORM Many platforms host an area for campaign updates, which are pushed out to your current backers. These are really important and show how dedicated you are to the goal. Plus, they make folks feel like insiders. Aside from updates, consider putting up a blog devoted to your project. If you have enough email addresses of interested people, you can also create an email list.

196 LEARN THE BASICS OF ROBOT ANATOMY

MAKE YOUR OWN

A robot is a so-called smart machine that does three basic things: sense, plan, and act. Various sensors collect data from the environment, while a computer analyzes this data and plans a course of action. Various actuators (motors, servos, arms, and grippers) are pressed into service in response. Let's look at the major robot subsystems and what your choices are for building your own 'bot.

A BRAINS You can use anything from cheap and simple microcontrollers (like the Arduino) to more robust single-board computers (like the Raspberry Pi), and from boards specifically designed for robots to full-fledged PCs. Which brain is dependent on your robot's design and control needs. A simple 'bot that navigates a space and has a few sensors needs only an Arduino or Raspberry Pi. But multiple sensors, video vision, and other bells and whistles need a computer that can handle them.

B SENSORS Robots are basically computers that move and interact with their world. To do this, they need devices that sense their environment. Numerous sensors on the market can detect touch, distance, motion, heat, light, and more. Some (e.g., infrared distance detection) generate analog values (continuously variable waveforms); others (e.g., contact switches) generate binary (on-off) values. Analog sensor data must be converted to digital using analog-to-digital hardware and software.

C POWER SYSTEM Your robot wants to roam, and that requires batteries. Tiny bugbots and other low-power robots can live off solar power, but photovoltaic and battery technologies are still limiting in many cases. As battery technology (slowly) advances, 'bot builders quickly embrace new innovations. The latest, lithium-polymer (LiPo) batteries, have an advantage over the older lithium-ion (Li-ion) batteries because they don't have liquid electrolytes (no leaking). LiPos can be formed into shapes, lower voltage/amp hour versions can be very thin, and they can pack high voltages in small volumes.

Li-ions have a few drawbacks (they don't age gracefully), but they're cheaper, widely available, and acceptable for many robot applications. Other formulations, such as nickel-metal hydride and sealed lead acid, are common in heavy-duty machines such as combat robots.

D ACTUATORS These are usually arms and end effectors (called grippers) driven by servomotors, sometimes legs for walking, and frequently wheeled locomotion orchestrated by motors and a drive train.

E DRIVE TRAIN Drive trains are the most common form of robot actuation, and usually include one or more drive motors, a gear train to deliver the right amount of motive power to the 'bot, and wheels or legs. Many basic 'bots—from quarter-size swarmbots to much larger development robots—employ two wheels, two motors, and two gear trains. The robot's brain issues commands to steer by changing the direction of rotation of the two wheels, similar to tank steering. Idle wheels or casters are used to keep the robot balanced.

F CHASSIS A robot chassis can be made out of many different materials. Miniature robots frequently employ light plastic. Some hobbyists add brains and servomotors to robot toys. Larger bots, combots, and industrial robots usually have metal frames. With the greater availability of CNC and 3D printers, many builders are printing their own robot parts. There are hundreds of robot parts, motor/servo mounts, and even 3D-printable robot bodies available for download on Thingiverse.com.

> "Finally, you can build your own mighty army of tiny fragile robots—all of your own devising."

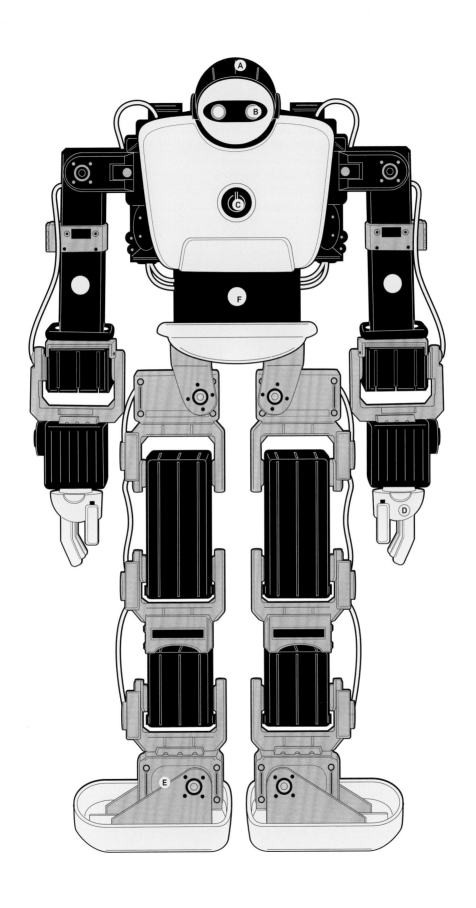

197 BUILD A SIMPLE PROXIMITY DETECTOR

One scheme for noncontact obstacle avoidance is infrared (IR) proximity detection. Here the idea is not to dynamically measure the distance to an obstacle but simply to initiate a behavior (obstacle avoidance) when the robot's distance from an object reaches a trigger value. Similar to a bump switch, when that proximity value is reached, it sends a signal to software to initiate the avoidance routine. A common, very simple circuit used to detect proximity is an infrared LED and an IR phototransistor. Similar to an ultrasonic sensor (though here we're measuring light, not sound), the LED sends out an IR signal that the phototransistor reads.

Follow the diagram in assembling your circuit. You may want to use a breadboard to test it out before soldering it onto your perfboard.

MATERIALS

- Infrared LED
- 24.7-ohm resistor
- Infrared phototransistor
- 10k-ohm resistor
- Heatshrink tubing (to fit over the phototransistor)
- Perfboard (to house your finished circuit)
- Breadboard (optional)

BUILD

STEP 1 Connect the IR LED to the 24.7-ohm resistor. The other end will eventually connect to 5 volts DC. The other LED pin goes to ground.

STEP 2 Connect the emitter pin on the phototransistor to ground, and then the collector pin (the longer pin) to the 10k-ohm resistor and then to the 5-volt supply. A wire is attached to the collector pin side of the phototransistor to serve as the output for the circuit.

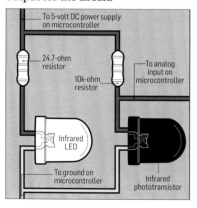

STEP 3 To make sure light waves from the IR emitter (or ambient reflected light) don't trigger false values, add some heatshrink tubing over the phototransistor receiver. **Pro tip:** If you don't have heatshrink (or electrical tape), you can just use cardstock or other material to make a blinder between the two bulbs.

STEP 4 To hook up your detector circuit, attach the 5-volt line to the 5-volt header on your microcontroller unit, the ground to the GND header, and the output pin/wire to an analog input on the microcontroller.

STEP 5 For the code you'll need, there are a number of Arduino sketches for this circuit already floating around the Internet. Do a search on "proximity sensor Arduino code." Install the code, run the program, and you should be good to go. **Note:** If you use an existing sketch, check to make sure it's using the same circuit (there are variations) and the input pin number is correct.

To alter the sensitivity of the circuit, change the resistor value on the phototransistor (lower will decrease sensitivity, while higher will increase sensitivity).

So, how does it work? Basically the emitter sends out infrared light waves that bounce off of a surface and back to the robot, and then get captured by the phototransistor receiver. The computer measures the light hitting the receiver and then uses that data to calculate a rough distance (and to trigger an avoidance behavior).

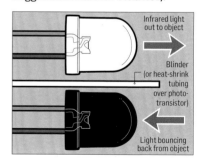

"Now that you've got your 'bot built out, time to endow it with the various gifts that you think best. It's kind of like playing God, only better."

198 MAKE YOUR ROBOT SEE

Adding vision to your bot—getting it to visually discern an environment, avoid objects, and recognize colors—is a little more difficult than some other senses, but there are many relatively inexpensive plug-in camera boards and free software that make it easier. For most microcontrollers, like the Arduino, you need a special board to interpret the video signal and hand that interpreted data to the MCU for processing. If your 'bot has a Raspberry Pi for a brain, it'll need a Raspberry Pi camera module.

ADD MORE SENSORS

- **HEAT** Build a fire-fighting robot.
- **TILT SWITCH** Know when your 'bot has flipped.
- **GPS** Add location awareness.
- **PIR SENSOR** Detect motion.
- **MICROPHONE** Give your robot hearing.
- **FORCE-SENSITIVE RESISTOR** Detect weight and pressure.

199 GIVE YOUR 'BOT TOUCH SENSE WITH GUITAR STRING

The most common and basic sense you can give your 'bot is touch. Just add one or more push-button switches to your 'bot, with bumpers or "whiskers" that will trigger the switch and initiate a reverse-and-turn routine in your software when in contact with an obstacle. With that simple sense, you have a robot that can navigate a space.

You can make an easy and highly effective touch sensor with little more than a guitar string (or similar wire), some heatshrink tubing, and a ring crimp connector. It's called a Scotty sensor, named after its inventor, Scott Martin.

To activate, simply connect one wire to anywhere on the ring connector and one to the end of the guitar string. When the string bends and touches the exposed metal of the crimp, it closes the switch. The plastic tubing insulates between the two parts of the connection. Easy! **Pro tip:** Add a metal sleeve inside the crimp connector (but not touching it) if you want to make your whiskers more sensitive.

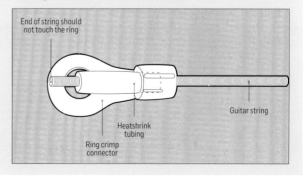

200 HELP YOUR 'BOT WITH DISTANCE DETECTION

Want your robot's room-navigation skills to be a little less touchy-feely than bumpers or whiskers? Then you want some sort of distance detection. A common form of this uses ultrasonic sound waves. The popular PING))) sensor from Parallax is a great choice for easily giving your 'bot ultrasonic superpowers. Basically, the small device (¾-by-1¾-by-½ inches [2-by-4.5-by-1.25 cm]) sends out an ultrasonic pulse (from 1 inch to 10 feet [2.5 cm–3 m]). From there, the software calculates distance from the round-trip time of that sonar ping. The PING))) is easily compatible with the BASIC Stamp, Arduino, Raspberry Pi, and other common microcontrollers.

201 HACK A SERVOMOTOR FOR CONTINUOUS ROTATION

A servomotor is a DC motor, gearbox, and motor controller circuit board all housed in a plastic or metal housing. The circuit board controls the range of motion of the motor and gears, usually in an arc limited to 130 to 180 degrees of motion.

Servos are used to precisely control back-and-forth movement, like in robot leg and arm, and torso joints. There is a vast array of servos available, from tiny quarter-size units to huge industrial models. Hobbyists often hack them, removing the PCB and modifying the main gear to create geared, continuous-rotation motors for robot drive systems. Here's how to remove the control electronics.

STEP 1 Put tape over the top case seam.

STEP 2 Remove the four screws from the case's bottom.

STEP 3 Desolder the two motor posts.

STEP 4 Remove the control PCB.

STEP 5 Solder new wires on as shown and replace the case bottom. Then, once the electronics have been removed, flip over, remove the tape, and remove the "final gear" (the one with the little tab-stop on it). Cut off that tab with a rotary tool or similar, carefully reseat the gears and rescrew the case together. You've now turned a servomotor into a continuous rotation gear motor.

   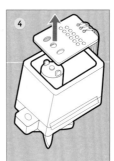 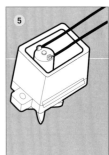

202 CLOWN AROUND WITH A BALLOON GRIPPER

Anyone who's played around with grippers and other finger-y end effectors for robots knows what a nightmare they can be to build and program. So why bother? The ingenious universal gripper (a.k.a. balloon gripper) offers surprising versatility, and it's made from little more than a balloon, coffee grounds, and a controllable suction source.

STEP 1 Put ground coffee into a rubber balloon using a small plastic funnel. Inflate the balloon a little to keep the grind well packed at the bottom. Keep filling it until your balloon fits comfortably into the mouth of the funnel.

STEP 2 Cut off some of the funnel throat to make a bigger opening. Feed the open end of the balloon through the funnel and secure it around the throat with duct tape.

STEP 3 Add a small wad of nondense fabric or fabric batting into the throat of the funnel and secure. This will prevent coffee from getting sucked out of the funnel.

STEP 4 Securely tape the balloon funnel gripper assembly to flexible air hosing and an air pump. That's it!

STEP 5 To grip an object, slightly inflate the balloon so that it deforms over the object you wish to pick up. Then suck the air out of the balloon so that it can firmly grip and pick up your object. Add computer control to do the blowing and sucking, and you have a universal robotic gripper that can handle far more diverse objects than fussier end effectors that are 10 times more complicated.

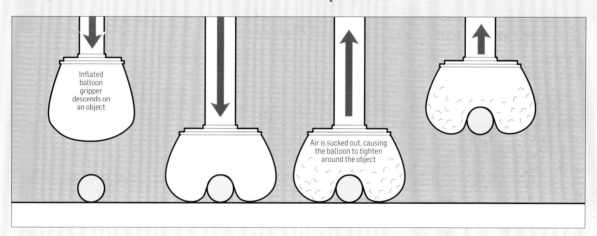

"Want your robot to be able to pick up after itself? Start with this simple balloon gripper—someday soon it will be crushing enemy skulls on its own."

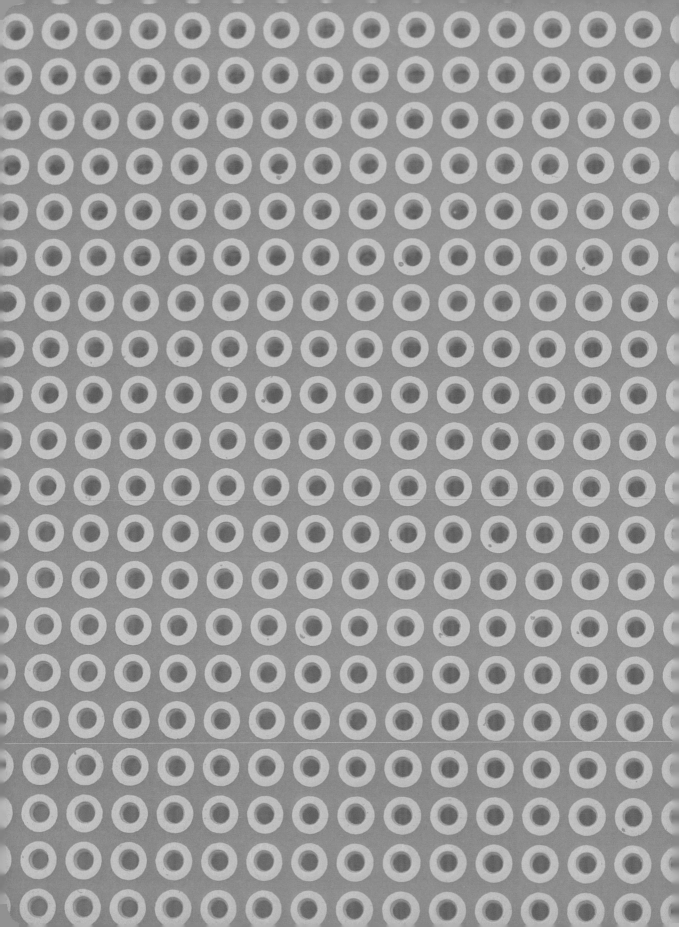

RESOURCES

GLOSSARY

3D PRINTING Also called additive manufacturing (AM). Refers to the mechanical production of three-dimensional objects from a digital file. Successive layers of material are laid down under the control of a 3D printer—a type of industrial robot.

AC (ALTERNATING CURRENT) An electrical charge that reverses direction at regular intervals.

AMP (AMPERE) A basic SI unit of electrical current that measures the amount of electric charge passing through a point in an electric circuit per unit time.

AMPLIFIER Component that augments the power of a signal. In circuits, an amplifier is usually used to increase the voltage or current.

ARDUINO Common, open-source microcontroller. There are various types of Arduino microcontrollers, but all can be programmed using the same programming language.

BANDSAW A power saw used in woodworking, consisting of an endless toothed steel band passing over two wheels.

BATTERY Two or more primary cells connected together that convert chemical energy into electrical energy.

BREADBOARD Base used to set up temporary circuits and test them out before soldering components together.

CAPACITOR Electrical component that stores energy within a circuit. Unlike a battery, a capacitor does not produce energy, it simply contains or filters the energy already flowing through the circuit.

CIRCUIT Closed loop through which electrical current flows. A circuit is often used to power an electrical device.

CIRCUIT BOARD Thin, insulated board on which electrical components are mounted and connected together. A printed circuit board (PCB) has thin conductive strips printed on the board, allowing connections to be made between components largely without the use of wires.

CIRCULAR SAW A power saw consisting of a toothed disc that uses rotary motion around a mandrel to cut different materials.

CLAMP Device used to hold an object tightly in place. Clamps can vary widely in size and construction, and can be intended for temporary or permanent use.

CNC MILLING A cutting process that uses automated machine tools to remove material from the surface of a base material. The mill is operated by programmed commands on a computer, as opposed to manually by hand wheels or levers.

COIN-CELL BATTERY Also called a button cell, a coin battery is a small, flat, disc-shaped battery that is often used to power portable electronic devices.

CONDUCTIVITY Capacity to transmit an electrical current; it can also refer to the measure of a substance's ability to transmit electrical current.

CONTACT Point where an electrical component is connected to a wire or circuit board.

DC (DIRECT CURRENT) An electrical current that flows in one direction only.

DESOLDERING Removing solder to detach components from a circuit or circuit board. Desoldering can be used to fix a fault in a circuit or to replace a component.

DIODE Electrical component with one terminal that has high resistance and another terminal with low resistance. A diode is used to allow current to flow in one direction but not another.

DRILL Tool used to cut holes in a variety of materials. A drill is usually powered by electricity and comes with an array of interchangeable bits in different sizes.

ELECTRICAL TAPE Type of tape covered in an insulating material, often used to cover and connect electrical wires.

ELECTRICAL WELDING A type of welding that uses an electrical current to create an electrical arc between an electrode and base material to melt metals at the welding point.

ELECTRICAL WIRE Insulated strand of conductive material used to carry electricity.

ELECTRODE Conductor used to transmit current to a nonmetallic material. Electrodes are used in arc welding to fuse objects together.

EPOXY Adhesive made from a type of resin that becomes rigid when heated or cured.

EXHAUST FAN Fan used to ventilate a work space; it is particularly important to use an exhaust fan when working with materials that emit toxic fumes.

FOUNDRY A manufacturing plant that produces metal castings. Smaller, more DIY versions can also be achieved.

GEAR A rotating, toothed machine part, typically a wheel or cylinder, that meshes with another toothed part to transmit torque in order to change the speed and direction of a larger machine.

GROUND WIRE Wire in a circuit that provides a return path for current, often leading to the earth. A ground wire can prevent the buildup of dangerous static electricity in a circuit.

HACKSAW Fine-toothed saw held in a frame. A hacksaw can be used to cut metal or other hard materials.

HEATSINK Device that channels heat away from an electronic system, keeping it cool enough to operate properly.

HOLE SAW Cylindrical saw blade used to cut holes of uniform size. A hole saw is usually used in a drill, in place of the drill bit, to create a large hole.

INSULATION Material, such as the nonconductive coating around an electrical wire, that prevents current or heat from flowing.

INTEGRATED CIRCUIT Also called a microchip, this small, thin device is a complete circuit etched or imprinted on a semiconductive surface. An integrated circuit allows for complex circuitry to be condensed into an extremely small space. It's vital to the operation of many modern electronic devices, notably computers.

JIGSAW Tool (usually a power tool) with a long, thin saw blade. A jigsaw is useful in cutting curves and irregular shapes.

JOINT The point at which two objects are connected together. In woodworking, creating a joint may involve cutting a notch or angle into the pieces of wood to be joined; in metalworking, the process often involves soldering or welding.

LASERCUTTING A technology that uses the output of a high-power laser to cut through materials.

LEAD A wire extending from an electrical component that is used to connect that component to another electronic part.

LED (LIGHT-EMITTING DIODE) Diode that gives off light. They are usually more energy efficient than incandescent light sources, and can be much smaller.

MICROCONTROLLER Tiny dedicated computer contained on a single chip that can be embedded within a larger device.

MULTIMETER Device that measures electrical current, resistance, and voltage. A multimeter is very helpful for monitoring and identifying problems in circuits.

OBTAINIUM Materials obtained secondhand, as salvage, by chance findings or donations. Used to create artwork or assemble structures that aren't bought new.

OHM Unit of measurement of electrical resistance.

OXYACETYLENE WELDING Also referred to as oxyfuel or gas welding, oxyacetylene welding is the process of cutting or welding metals by using a focused flame whose temperature is intensified by a mixture of fuel gases (acetylene) and oxygen.

PARALLEL CIRCUIT A closed circuit in which all the components are connected between two sets of common points, dividing the current into two or more paths before recombining.

PHOTOCELL Device that produces a flow of current when exposed to light. A photocell can detect the presence (or absence) of light or other radiation.

PHOTODETECTOR Any of various kinds of sensors that detect and measure the intensity of light or other radiant energy, such as photoresistors, photovoltaic cells, photomultipliers, and phototransistors.

GLOSSARY

POTENTIOMETER Three-terminal electrical component that acts as a variable resistor. These adjust the flow of current through a circuit and are often used in dimmer switches or volume controls.

PROGRAMMING LANGUAGE Language used to convey instructions to a computer or other machine. Many distinct programming languages are used for different purposes and types of hardware.

REBAR Ridged bar of steel often used in construction to reinforce concrete or masonry.

RECIPROCATING SAW A power tool that cuts with the back-and-forth motion of a saw blade.

RESISTOR Two-terminal electrical component that resists the flow of an electrical current. A resistor is used in a circuit to control the direction and strength of the current flowing through it.

ROTARY TOOL Power tool with a wide variety of interchangeable bits that can be used for different purposes. A rotary tool can cut, polish, carve, or grind; it's particularly good for detail work.

SAFETY GOGGLES Glasses that shield the eye area from heat, chemicals, and debris.

SCHEMATIC Two-dimensional map of an electrical circuit. A schematic uses a set of symbols to stand for the components of a circuit and shows the connections between them.

SENSOR A device that receives stimuli, or detects changes in quantities in its surrounding environment, and responds with a corresponding output, typically as an electrical or optical signal.

SERIES CIRCUIT A circuit with all the resistors arranged in a row, so that there is only one path for the current to flow.

SKETCH When working with Arduino microcontrollers, a sketch is a program that can be loaded into an Arduino.

SLAG Scrap metal that results from the welding process.

SOLDERING Connecting two metal objects together by melting solder (a type of metal) with a soldering iron to create a strong joint between the objects.

SWITCH Component that can stop the flow of current in a circuit, or allow it to continue. These include push-button switches, rocker switches, toggle switches, and many other devices.

TABLE SAW Machine that cuts wood or other materials with a rapidly spinning serrated metal disc. A table saw is usually powered by electricity and often mounted within a safety guard.

TRANSFORMER Device that transfers current from one circuit to another. A transformer can also be used to alter the voltage of an alternating current.

TRANSISTOR Semiconducting electrical component with at least three leads; can control or amplify the flow of electricity in a circuit.

VISE Type of clamp, often affixed to a table, that uses a screw to hold an object tightly in place.

VOLT Unit of measurement for electrical potential.

WELDING Process of joining pieces of metal together by melting them slightly and introducing a filler material at the joint.

WIRE CUTTER Pliers with sharp diagonal edges used to cut lengths of wire.

WIRE STRIPPER Device composed of a set of scissorlike blades with a central notch; used to strip the insulation from the outside of electrical wires.

ZIP TIES Self-closing fastener. When the end of a zip tie is inserted into the slot at its head and tightened, it creates a loop that can't easily be loosened.

INDEX

A
AC (alternating current) 119, 122, 156
AC power switch 185
AC-powered rotary tool 102
active speakers 151
Adafruit Flora 181
add-on board 187
addressable LEDs 185
adhesives 75
air rotary tool 102
air sensor 185
alphanumeric LEDs 126
aluminum 64, 104
amplifier 152
anchors 20
AngelList 193
angle grinder 92, 93–95
angle vise 42
Arduino microprocessors 127, 181, 184, 187, 196–198, 200
Arduino Uno 181, 187
Arduino Yún 181
audio circuits 152

B
backstitch 82
baling wire 42
ball peen hammer 18
balloon gripper 202
bandsaws 85, 88–91
battery chargers 165
battery-powered rotary tool 102
Beaglebone Black 181
bench belt sander 93
bench grinder 93
bench mount 90
bench power supply 165
bench saw 85
bench vise 42
bicolor LEDs 126
bicycle generator 171–172
blacksmithing 56
blinking LED 127, 184
bolts 20
bottles, how to cut 72
box wrench 27
brainwave sensor 185
brazing 50
breadboard 117

bricks 65, 68
Bristol fastener head 22
butt joints 36
buzzers 150–151

C
C clamp 42
cable crimps 73
CAD (computer-aided design) software 174, 178
caliper 13
CAM (computer-aided manufacturing) software 174, 178
capacitor 119, 122, 131–133, 150, 152, 161
cardboard 77
Cartesian robot 175
casting 58, 73
catch stitch 82
cell-phone blocker 158
cement 65
chain-drilling 100
cheater bar 29
Chicago screws 84
chisels 41
cinder blocks 65
CircleUP 193
circuit bending 155
circuits 116–119, 121, 134, 139, 145, 146, 152
circular saw 85, 87, 88
clamps 42
class-AB power amp 152
claw hammer 18
clear packing tape 74
clockmakers' hammer 18
CNC (computer numerical control) 173–175, 192, 198
CNC routers 173, 175
coil/choke 119
coin-cell batteries 165
combination square 12
combination wrench 27, 30
concrete 65, 66, 67
conditioning disc 95
conductive fabrics and threads 139–145
continuity check 136
coping saw 33
copper 55, 104, 118, 119, 159
corner clamp 42
countersink bit 101
crescent wrench 27

crosscut saw 33
Crowdfunder 193
crowdfunding campaign 193–195
crucible carrier 61
crucible, how to build 60
crucible pourer 62
crush injuries 8
current:
   AC 156
   DC 156
   inverter 172
   measuring 138
   redirecting 124
cutoff wheel 96
cuts, first aid for 6
cyanoacrylate 75

D
DC (direct current) 156
DC welding 111, 112
deadblow hammer 18
decibels 149
dial caliper 13
die grinder 93, 100, 102
digital caliper 13
discs and wheels 95, 96
DIY amp 152
DIY welder 111, 112
dowels 20
DPDT (double-pole double-throw) switch 160
DPST (double-pole single-throw) relay 124
Dremel 102
drill bits 101
drill parts, salvageable 162
drill press, improvised 103
drilling 97–100
dross skimmer 63
duct tape 74
dynamic microphones 153

E
ear protection 5
electret microphones 153
electrical welding 105–112
electronic circuits 116–118
electronic components 119
EquityNet 193
etching 118, 173, 174, 189–191
eye flush 7
eyebolts 20

F

fabric 139–142
fading an LED 127
fastener heads 22
fasteners 20–26, 84
FFF (fused-filament fabrication) 176, 178
fire suppression 46
first aid 5–9
first aid kit 5
555 IC timer 131–134
flame-cutting 53, 54
flasher, free-running 132
flashing LED 126
flashlight, solar-powered 169
flat-head screws 22, 24
flat-nose pliers 73
flex sensor 143, 185
flexible shaft 102
flexible switches and sensors 139, 143
flooring for work spaces 46, 47, 77
fluorescent bulb 126
forge:
   how to build 56
   projects 57–58
foundry, how to build 59
fuel cells 165
fused-filament fabrication (FFF) 176, 178

G

gaffer tape 74
gas tank hose leaks 48
gas tanks 46
gas welding 46–64
gearbox, shimming 164
gears 163–164
generator 11, 172
Ghazala, Reed 155
glass 65, 70–72, 98
glass drill bit 101
glass mounting 71
glass work space setup 70
glues 42, 75
grinders and sanders 93–96
grinding discs 95
grommets 84

H

hackerspace 175, 177, 192
"Hackett says" 19, 43, 92, 110, 166, 179

hacksaw 33
halogen bulb 126
hammers 14–18
hand jigsaw 33
hand tools 10–45, 10–84
handheld belt sander 93
hanger wire 42
hardware 1
hardwood 34
hearing protection 5
hex fastener head 22
hole saw 33, 56, 59, 99, 101
homopolar roller 159
hose leaks 48
HPLED (high-power LED) 129
HVAC ducts 40, 74

I

illumination sources 126
image sensors 146
incandescent bulb 126
Indiegogo 193
infrared lights 126
injuries 6–9
integrated circuit 119
Intel Galileo 181
inverter 172
iron 104

J

Japanese flush-cutting saw 33
jewelers' hammer 18
jewelers' saw 33
jigs 42
jigsaw 85, 86

K

keyhole saw 33
Kickstarter 193
knife, forged 57
knock sensor 153

L

lasercutting 188–191
LDR (light-dependent resistor) 150
LED strips 126
LED types 126
LEDs 120–121, 124–133, 137–138, 141, 146, 184, 185
leverage 29
lightbulb types 126
lighting 1
linear motor 185
linear softpot 143
lineman's pliers 73
LM386 chip 152

lockjaw pliers 73
Loctite 76
loop switch 148
lumber numbers 34

M

machinist vise 42
makerspace setup 175, 177, 188, 192
mallet 18
masking tape 74
masonry 65–72
masonry bit 101
masonry chisels 41
maul 18
measuring tools 10–13
mechanical advantage 19
metal:
    bending 51
    brazing 50
    cutting with snips 40
    description 104
    flame-cutting 53, 54
    welding 49
metal chisels 41
micro measurements 13
microcontroller accessories 182
microcontroller add-on board 187
microcontrollers 181–185
microphones 153–154
MIG welding 105, 106, 108
milling machine, improvised 103
Mims, Forrest 147
miniature LEDs 126
monkey wrench 27
motors 159–162
moving heavy objects 16
multimeter 135–138, 142, 144, 161
multiplexing 130

N

nails 20
nailset 17

needle-nose pliers 73
no-sew fasteners 84
nuts 20

O

obtainium 55
Ohm's Law 125
oil finish 35
one-shot timer 133
one way fastener head 22
orbital sander 93
oscillator, free-running 132
overcast stitch 82
oxyacetylene torches 46
oxyacetylene weld 49
oxyacetylene welding 46–54

P

paint-can forge 56
paper 77
Parallax BASIC Stamp 181
Parallax PING))) sensor 200
Parallax Propeller 181
parallel wiring 128
passive cone speakers 151
pattern-making 83
PCB mills 173, 175
PCB (printed circuit board) 114
pedal power 109, 171
Peltier junction 165
person-powered generator 165
persuasion 18, 19
phase converter 165
Phillips fastener head 22
photocell 146, 150
photodetectors 146
photodiodes 146
photomultipliers 146
photoresistors 146
phototransistors 146
piezo buzzer 151
piezo crystal 185

piezo microphone 151, 153, 154
piezo speaker 133
pipe:
    cutting 77
    fishmouth technique 77
    threading 45
pipe clamp 42
pipe hangar 42
pipe wrench 27, 31, 32
pitch video 194
plaster 69
plaster mask 69
plastics 180
pliers 73
plumbing dies 45
portable bandsaw 85, 88, 89, 90, 91
Posidrive fastener head 22
potentiometer 119, 123, 143, 150, 155
power 1, 138
power, measuring 138
power saws 85–91
power sources 165–172
power strips 1
power tools 85–103
pressure sensor 147
pressure-treated lumber 34
printed circuit boards (PCBs) 114
programming primer 186
project communication 3, 5
project management 2–4
project planning 2, 77
project teams 3, 4
prototyping case 183
proximity detector 197
PVC fabric 142
PWM (pulse-width modulation) 127

R

radio 156–158
Raspberry Pi 181, 187, 196, 198, 200
reamer 101
reciprocating saw 85
regulators 46, 48
relays 124
repurposed metal 55
resistors:
    calculation 125
    description 119
    and high-power LEDs 129
    light-dependent 150
    photoresistors 146
    use in timers 131–133
    variable 123
reverse switch 160
RFID (radio frequency identification tag) 185
RGB LEDs 126
rheostat 123
ripsaw 33

robot anatomy 196
robot gripper 202
robot motion 201
robot sensors 197–200
robot tools 173–175
roll-up tool holder 142
rotary tools 93, 102–103
rubber cement 75
running stitch 82

S

safety goggles 46
sand paper 37
sander and grinders 93–96
sanding disc 95
sawing 38
saws 34
schematics 116
scrap metal 55, 104
screen printing 78–80
screw eyes 20
screw heads 22
screw posts 84
screws 20–26, 30
sensors 119, 139, 143, 146–148, 154, 181, 183, 185, 196–200
series wiring 128
servomotor 185, 196, 201
severed body part 9
sewing 81–84
sewing kit 81
sewing stitches 82
sex bolts 84
shearing 38
sheet metal snips 40
sheetrock saw 33
shop safety 5
shot glass, cast 58
SLA (stereolithography) 176
sledgehammer 18
slipjoint pliers 73
slots, how to drill 100
slotted fastener head 22
small drivers 151
snaps 84
snips, metal 40
socket wrench 27
soft circuit cuff 145
soft push button 143
softwood 33, 34
solar cells 146, 167, 168
solar cooker 167
solar panels 165, 168, 169
solar power 167–169
soldering 113–115
soldering station 113
spade bit 101
spanner 27, 28, 30

sparker lighter 46, 53
sparks 46, 47, 94, 104
speakers 151
specialty screw conversion 24
SPST (single-pole single throw) relays 124
square, combination 12
stainless-steel thread 141
steel 104
step drill bit 101
stepper motor 119
stereolithography (SLA) 176
stick welding 106, 107
Stillson wrench 31, 32
storage 1
stripped screw, removal 25
superglue 75
switches 119, 160

T

tack hammer 18
tape measurer 10, 11
tapes 42, 74
taps 43, 44
templates 77
textiles 83, 139–142, 189
thermoelectric element 185
thin speakers 151
threaded fasteners 20–26, 30, 76
threading a pipe 45
3D printing 175–179
TI Launchpad 181
TIG welding 105, 106, 109
timers 131–134
torque 26, 29
torque wrench 27
Torx fastener head 22
touch sensors 139, 143
transformer 119
transistor 119
trash 1
trench radio 157
tricolor/RGB LEDs 126
tri-wing fastener head 22
trimmer 123
tube cutter 39
twist drill bit 101
two-part epoxies 75

U

ultrasonic range finder 185
unthreaded fasteners 20, 21

V

ventilation 5, 46
vices 42
vinyl 142
vinyl tape 74
voltage, measuring 137

W

washers 20
water-based glues 75
welding:
  DC (battery-operated) 111–112
  electrical 105–109
  gas 46–54
welding filler metal 46
welding hoses 46, 48, 51
welding safety 110
welding space setup 47, 52, 105
welding station 46
welding table 52
wheels and discs 95, 96
wind turbine 170
wire saw 33
wire strippers 73
wire wheel 95
wires:
  how to solder 115
  types 42, 119
wiring 128
wood 34
wood chisels 41
wood finishes 35
work space set-up:
  general 1, 77
  for glass working 70
  for hackerspace 192
  for soldering 113
  for welding 47, 52, 105
worktable 1
wrenches 27–32

Z

zip ties 42

Suite 100
San Francisco, CA 94111
Telephone: 415 291 0100
Fax: 415 291 8841
www.weldonowen.com

PRESIDENT, CEO Terry Newell
VP, SALES Amy Kaneko
VP, PUBLISHER Roger Shaw
DIRECTOR OF FINANCE Philip Paulick

SENIOR EDITOR Lucie Parker
PROJECT EDITOR Goli Mohammadi
CONTRIBUTING EDITOR Nic Albert
EDITORIAL ASSISTANT Jaime Alfaro

CREATIVE DIRECTOR Kelly Booth
ART DIRECTOR Lorraine Rath
DESIGNER Allister Fein
ILLUSTRATION COORDINATOR Conor Buckley
SENIOR PRODUCTION DESIGNER
    Rachel Lopez Metzger

PRODUCTION DIRECTOR Chris Hemesath
ASSOCIATE PRODUCTION DIRECTOR
    Michelle Duggan

Copyright © 2014 Weldon Owen Inc.
All rights reserved, including the right
of reproduction in whole or in part in
any form.

Library of Congress Control Number: 2014950287

Paperback edition: 978-1-61628-726-9
Hardcover edition: 978-1-61628-890-7

10 9 8 7 6 5 4 3 2 1
2014 2015 2016 2017 2018

Printed in Canada

Popular Science and Weldon Owen are divisions of
BONNIER

PUBLISHER ACKNOWLEDGMENTS
Weldon Owen would like to thank Jan Hughes, Marianna Monaco, Katharine Moore, and Marisa Solís for editorial assistance, as well as our technical editor, Sean Michael Ragan.

In addition, we would like to acknowledge Valeryia Fateycheva, Jamie Spinello, and Brandi Valenza for storyboarding expertise, and photographer Christopher Beauchamp and photo assistant Michelle Chappel for original imagery.

We would also like to thank Dave Mosher, Sophie Bushwick, and Thomas Payne at *Popular Science*.

AUTHOR ACKNOWLEDGMENTS
I am all about DIY, but the term is a little misleading: You need other people to learn from, to keep you on track, and as a source of inspiration and ideas to steal.

This is true of this book—it wouldn't have happened without people pitching in, helping me move forward, calling me on my bullshit. These people get all of the credit and none of the blame: Aaron Klinglehoeffer, Lis Thomas, David Mosher, Maggie Sullivan, Julia Solis, Christos, Becky Stern, Sarah Edith, Gene Wayda, Erin McAdams, Ryan O'Connor, Mr Nice Guy, Sarah Sundberg, Matthew Griffin, Mike O'Toole, Chuck Messer, Colin Butgereit, Bonnie Downing, Jim, Kim!, Warbeast, Pat, Nina, Alex, everyone at the Madagascar Institute, McMaster Carr, the guys at the junkyard on 3rd Avenue and 25th Street next to the slaughterhouse, and you, special you: I might have left your name out—you can think of reasons why, but you know this part is meant for you.

Special thanks to Anna Cerminara, assistant of pluck to chaotic employer; extra special credit to Bronwen Densmore and Lucie Parker, who bore the stress and pain I doled out with grace and calm—if I wind up dead, chances are one of them did it, but they earned the right and should be lauded for waiting so long.

Last, I would like to thank ShopKat, without whom no good things may exist.

ADDITIONAL TEXT CREDITS

All makers could stand a little help from their friends. We'd like to thank the various tinkerers, builders, and creatives for lending wisdom and words to this book:

Gareth Branwyn: 196–202 **Abe Connally:** 170 **Blaine Dehmlow:** 188–191 **Sam Freeman:** 127–130, 149–154 **Matthew Griffin:** 176–178 **Steve Hoefer:** 181–187 **Goli Mohammadi:** 142, 193–195, 155 **Ryan O'Connor:** 35–37 **Syuzi Pakhchyan:** 139–141, 143–145 **Sean Michael Ragan:** 72, 122–124, 131–134, 146–148, 156, 159–164, 167–169, 173–175 **Lis Chere Thomas:** 81–84

PHOTO CREDITS

All photographs courtesy of Shutterstock Images unless listed below.

Back cover: Christopher Beauchamp

Adafruit Industries: 126 (LED strip, tricolor/RGB LED, bicolor/ RGB LED, infrared LED), 151 (piezo buzzer, thin speaker), 176 (flex sensor, addressable LED, thermoelectric element, AC power switch), 181 (flora, BeagleBone Black) **American Rotary:** 165 (phase converter) **Arduino:** 181 (Uno, Yún) **Armstrong Industrial Handtools:** 18 (deadblow) **Christopher Beauchamp:** 19, 43, 92, 110, 166, 179 **Beflux Electronics:** 126 (alphanumeric LED) **Bessy Tools North America:** 42 (corner clamp, pipe clamp) **Courtesy of Carnegie Mellon Robotics Academy:** 146 (photoresistor) **Jason Chinn:** 120 ("Aurora" installation by Charles Gadeken) **Creative Commons:** 101 (reamer), 146 (photomultiplier) **Dasco Pro:** 41 (masonry chisel, bottom) **Defense.gov News Photos:** 88 (portable bandsaw) **DeWalt:** 85 (portable bandsaw), 93 (orbital sander) **Dremel:** 93 (rotary tool), 101 (glass drill bit) **Elenco:** 165 (bench power supply) **Ed Gerrel:** 61 (dross skimmer) **Sam Freeman:** 130 **Genelec:** 151 (active speakers) **Gil C. / Shutterstock Images:** 193 **Grove:** 185 (air sensor) **Harbor Freight Tools:** 42 (baling wire) **Courtesy of The Home Depot:** 41 (masonry chisel, top two), 42 (pipe hanger), 85 (bench saw) **Intel Corporation:** 181 (Galileo) **Jet Equipment & Tools:** 27 (spanner wrench) **K-TOR:** 165 (pedal crank) **KEO Milling Cutters:** 101 (countersink drill bit) **Kingbright Electronics:** 126 (miniature LED) **Kondo:** 190 **Ray Lego:** 56 **Frits Lyneborg:** 153 (piezo microphone) **MakerBot:** 177 **NeuroSky:** 185 (brainwave sensor) **Matthew Newlove:** 155 **Syuzi Pakhchyan:** 139–140, 145 **Parallax Inc:** 119 (sensor), 181 (propeller, BASIC Stamp), 185 (servomotor, GPS, ultrasonic range finder) **ProtoStack:** 153 (electret microphone) **Sean Michael Ragan:** 72 **Raspberry Pi Foundation:** 181 (Raspberry Pi) **Rikon:** 93 (bench belt sander) **Rothco:** 33 (wire saw) **Sennheiser:** 153 (dynamic microphone) **Smithy Industries:** 42 (angle vise) **Julia Solis:** 55 **Speedway Motors:** 185 (linear motor) **Stanley Tools:** 33 (sheetrock saw) **Becky Stern:** 170 **Texas Instruments:** 181 (launchpad) **Tormach:** 42 (machinist vise) **UltraCell:** 165 (fuel cell) **Veritas Tools:** 33 (Japanese flush-cut saw) **Vollong Electronics:** 126 (high-power LED) **Wilton Tools:** 42 (bench vise)

ILLUSTRATION CREDITS

Conor Buckley: 6, 12, 14–15, 22, 52, 56, 82, 90, 106 (circuit), 111, 116, 121, 125, 127, 128, 131–133, 137, 143, 145, 148, 152, 161, 169 (circuit), 170, 172, 196–197 **Hayden Foell:** 26, 29, 44, 59, 106 (tools), 122–123, 150, 157, 176 **Vic Kulihin:** 30, 32, 45, 49, 61–62, 76, 94, 164, 187 **Raymond Larrett:** 28, 50, 53, 114–115 **Liberum Donum:** 1, 46, 113, 192 **Christine Meighan:** 54, 57–58, 60, 78–79, 86, 91, 98, 100, 103, 136, 160, 162, 184, 201 **Tina Cash Walsh:** 24–25, 38–39, 48, 71, 83, 112, 129, 147, 154, 167, 169, 199, 202

DISCLAIMER

The information in this book is presented for an adult audience and for entertainment value only. While the advice in this book has been fact-checked and, where possible, field-tested, much of this information is speculative and situation-dependent. The publisher assumes no responsibility for any errors or omissions and makes no warranty, express or implied, that the information included in this book is appropriate for every individual, situation, or purpose. Before attempting any activity outlined in these pages, make sure you are aware of your own limitations and have adequately researched all applicable risks. This book is not intended to replace professional advice from experts in electronics, woodworking, metalworking, or any other field. Always follow all manufacturers' instructions when using the equipment featured in this book. If the manufacturer of your equipment does not recommend using it in the fashion depicted in these pages, you should comply with the manufacturer's recommendations. You assume the risk and full responsibility for all of your actions, and the publisher will not be held responsible for any loss or damage of any sort—whether consequential, incidental, special, or otherwise—that may result. Otherwise, have fun.